FROM GENETICS TO MATHEMATICS

SERIES ON ADVANCES IN MATHEMATICS FOR APPLIED SCIENCES

Published:*

Vol. 65 Mathematical Methods for the Natural and Engineering Sciences
by R. E. Mickens

Vol. 66 Advanced Mathematical and Computational Tools in Metrology VI
eds. P. Ciarlini et al.

Vol. 67 Computational Methods for PDE in Mechanics
by B. D'Acunto

Vol. 68 Differential Equations, Bifurcations, and Chaos in Economics
by W. B. Zhang

Vol. 69 Applied and Industrial Mathematics in Italy
eds. M. Primicerio, R. Spigler and V. Valente

Vol. 70 Multigroup Equations for the Description of the Particle Transport in Semiconductors
by M. Galler

Vol. 71 Dissipative Phase Transitions
eds. P. Colli, N. Kenmochi and J. Sprekels

Vol. 72 Advanced Mathematical and Computational Tools in Metrology VII
eds. P. Ciarlini et al.

Vol. 73 Introduction to Computational Neurobiology and Clustering
by B. Tirozzi, D. Bianchi and E. Ferraro

Vol. 74 Wavelet and Wave Analysis as Applied to Materials with Micro or Nanostructure
by C. Cattani and J. Rushchitsky

Vol. 75 Applied and Industrial Mathematics in Italy II
eds. V. Cutello et al.

Vol. 76 Geometric Control and Nonsmooth Analysis
eds. F. Ancona et al.

Vol. 77 Continuum Thermodynamics
by K. Wilmanski

Vol. 78 Advanced Mathematical and Computational Tools in Metrology and Testing
eds. F. Pavese et al.

Vol. 79 From Genetics to Mathematics
eds. M. Lachowicz and J. Miękisz

*To view the complete list of the published volumes in the series, please visit:
http://www.worldscibooks.com/series/samas_series.shtml

Series on Advances in Mathematics for Applied Sciences – Vol. 79

FROM GENETICS TO MATHEMATICS

Edited by

Mirosław Lachowicz
Jacek Miękisz

University of Warsaw, Poland

World Scientific

NEW JERSEY • LONDON • SINGAPORE • BEIJING • SHANGHAI • HONG KONG • TAIPEI • CHENNAI

Published by

World Scientific Publishing Co. Pte. Ltd.
5 Toh Tuck Link, Singapore 596224
USA office: 27 Warren Street, Suite 401-402, Hackensack, NJ 07601
UK office: 57 Shelton Street, Covent Garden, London WC2H 9HE

British Library Cataloguing-in-Publication Data
A catalogue record for this book is available from the British Library.

FROM GENETICS TO MATHEMATICS
Series on Advances in Mathematics for Applied Sciences — Vol. 79

ISBN-13 978-981-283-724-0
ISBN-10 981-283-724-8

Printed in Singapore.

Stanisław Cebrat

Ryszard Rudnicki

Mirosław Dudek

Adam Lipowski

Preface

Recently an intensive development of mathematical methods in biological sciences and in particular in genetics is observed. From one point of view, one needs new statistical methods and algorithms to classify and interpret a huge collection of data. On the other hand, there is a need for analytical mathematical methods helping to understand physical and biochemical processes on the molecular level [1].

Modern genetics uses a wide spectrum of mathematical models which describe gene distributions in evolving populations, changes in a single genome or biochemical processes regulated by genes, *cf.* [2, 3] and references therein. Although genetics is concerned with small biological objects, in some cases methods of population dynamics can be successfully applied to study genetic problems [4]. Most models are formulated on the macroscopic level and the possibility of modeling on the level of basic objects involved in processes is still open [5].

This volume contains review papers based on two short courses given by a genetist Stanisław Cebrat and a mathematician Ryszard Rudnicki and two lectures given by physicists, Mirosław Dudek and Adam Lipowski. The courses and lectures were presented during the school "*From Genetics to Mathematics*" held in Zbąszyń, Poland, 18–20 October 2007 (see [6]), organized in the framework of the Marie Curie Research Training Network Research MRTN-CT-2004-503661 "*Modeling, Mathematical Methods and Computer Simulations of Tumor Growth and Therapy*". The goal of the school was to gather genetists, mathematicians, and physicists to discuss recent advances in mathematical modeling of biological problems in genetics and discuss possibilities of future collaborations. In particular, the school was dealing with issues and problems in genetics which could be addressed by mathematics and theoretical physics. All talks were of expository nature and had a strong pedagogical character and such are papers included in this volume.

The first three chapters are provided by Stanisław Cebrat and his coworkers. They present an extensive introduction to genetics and discussed their own ideas and models of evolutionary genetics. In particular, evolution of coding sequences, evolution of whole genomes, the noisy Penna model, sympatric speciation and recombination effects are discussed. These dynamical models, like the Penna model, were investigated so far only by numerical simulations. Mathematical analysis of such models may provide some further intuitions and new results.

The aim of Chap. 4, written by Ryszard Rudnicki, is to give a survey of mathematical models and methods of population dynamics in application to genetics. Both discrete and continuous in time and structure models are considered. The author provides the historical background of mathematical modeling of population dynamics, including the classical Lotka–Volterra model and the Kermack–McKendrick model of epidemics. He discusses the Penna model, the model of the evolution of paralog families in a genome, cell-cycle models, and a model of stochastic gene expression. This chapter shows that fundamental models in genetics lead to deep mathematical problems in dynamical systems, partial differential equations, and the theory of Markov operators.

Lotka–Volterra–type models are discussed from different points of view in Chaps. 5 and 6.

Mirosław Dudek and Tadeusz Nadzieja analyze the growth of age–structured population with genetics. They discuss time–delay growth equations with age structure, Lotka–Volterra models with genetics and present computer simulations of the Penna model.

Adam Lipowski and Danuta Lipowska introduce spatial Lotka–Volterra models. To investigate long–run behavior of finite systems of many interacting objects, one has to take into account stochastic fluctuations. They discuss two–species and multi–species models of ecosystems with a particular emphasis on the problem of extinctions. They introduce also a new model of the naming game of bio–linguistics.

We gratefully acknowledge the help of dr Monika Piotrowska, without her the book would not have been completed. We would like to thank Katarzyna Kutzmann–Solarek and Ireneusz Solarek from Zbąszyń Culture Center for their help and the possibility of participation in the performance "3 states of matter" during the school in Zbąszyń. Last but not least we thank Marian Kwaśny, the blacksmith from Nądnia, for the boat cruise in the Błędno lake.

Mirosław Lachowicz and Jacek Miękisz

References

[1] R. Bürger, C. Maes, and J. Miękisz, Eds., *Stochastic Models in Biological Sciences.* (Banach Center Publications 80, Warszawa, 2008).
[2] A. W. F. Edwards, *Foundations of Mathematical Genetics.* (Cambridge University Press, Cambridge, 2000).
[3] W. J. Ewens, *Mathematical Population Genetics I: Theoretical Introduction.* (Springer, 2004).
[4] R. Rudnicki, Ed., *Mathematical Modelling of Population Dynamics.* (Banach Center Publications 63, Warszawa, 2004).
[5] J. Banasiak, V. Capasso, M. Chaplain, M. Lachowicz, and J. Miękisz, in *Multiscale Problems in the Life Sciences. From Microscopic to Macroscopic,* V. Capasso and M. Lachowicz, Eds., Lecture Notes in Mathematics 1940 (Springer, 2008).
[6] `http://www.mimuw.edu.pl/~biolmat/zbaszyn.html`

References

[1] [illegible]

[2] A. [illegible] 2007.

[3] [illegible]

[4] [illegible] Topological Dynamics [illegible]

[5] [illegible]

[6] [illegible]

Contents

Chapter 1

To understand nature - computer modeling between genetics and evolution

Dorota Mackiewicz and Stanisław Cebrat

Department of Genomics, Faculty of Biotechnology,
University of Wrocław,
ul. Przybyszewskiego 63/77, 51-148 Wrocław, Poland,
cebrat@smorfland.uni.wroc.pl.

We present the basic knowledge on the structure of molecules coding the genetic information, mechanisms of transfer of this information from DNA to proteins and phenomena connected with replication of DNA. In particular, we describe the differences of mutational pressure connected with replication of the leading and lagging DNA strands. We show how the asymmetric replication of DNA affects the structure of genomes, positions of genes, their function and amino acid composition. Results of Monte-Carlo simulations of evolution of protein coding sequences show a specific role of genetic code in minimizing the effect of nucleotide substitutions on the amino acid composition of proteins. The results of simulations are compared with the results of analyses of genomic and proteomic data bases. This chapter is considered as an introduction to further chapters where chromosomes with genes represented by nucleotide sequences were replaced by bitstrings with single bits representing genes.

Contents

1.1. Introduction

> One day, Leo Szilard informed his friend that he is going to write a diary. "I am not going to publish it but just to inform God about some fact". "Do you think that God doesn't know facts?" asked his friend. "I am sure, God doesn't know this version of facts" answered Leo.

This chapter is a kind of the diary. It is written in a few months but authors try to show their way of understanding the Nature or rather of understanding some aspects of the biological evolution. "Their way" means both - how they understand and what was the chronology of understanding the facts.

In the first part, the simplest rules of coding the genetic information are described. Everybody knows what is the genetic code, everybody knows what is the structure of the DNA double helix, maybe even how it is translated for amino acid sequences in proteins, but it is not a common knowledge how Nature exploits the fact that the two strands of DNA differ in the mechanism of their synthesis, composition, mutation pressure and selection. It is the only part where genes are represented by long series of different elements which have to be translated into chains of other compounds and their biological functions are checked at the end of the process of the virtual evolution.

In the next parts these genes with very complicated structures are replaced by single bits. Reader can intuitively feel that it is a swindle. Nevertheless, remember that it is not a physicist who tries to convince you that such simplification is possible - it is a biologist. Physicists are very skillful in such simplifications, they were even able to count the planets paths assuming that they are just points - everybody knows that Earth is not a point! Using our simplification that gene is represented by one bit we were able to answer such questions like why men are altruistic and women are egoistic, why women live longer than men, why Y chromosome is short, why Nature invented death and if there is any significant difference between the fruit flies and humans. All answers have been obtained assuming that at least a part of genomic information is switched on chronologically.

In the third part we have even resigned from this chronology, instead, we have introduced some specific behavior of our virtual creatures, particularly their sexual behavior. Using a very simplified mechanism mimicking meiosis (unfortunately necessary in sexual reproduction as well as in the modeling such reproduction) we were able to show that speciation is a very simple phenomenon even in sympatry. The last results would be interesting even for Darwin himself when writing his "The Origin of Species by means of Natural Selection".

Since it is a biological paper, rather, and the mathematical formal language is too difficult for biologists, you should not expect formulas in the text. But the text is dedicated for mathematicians and they simply can translate the results obtained by the Monte-Carlo methods for their beautiful phenomenological language. I hope that everybody, after reading the text will have much more questions and problems which could be quantitatively described, and readers should judge if God should be informed about such quantitative versions of the biological facts.

1.2. Evolution of the DNA coding sequences

1.2.1. *DNA double helix*

Before Watson and Crick discovered the DNA (DeoxyriboNucleic Acid) structure, Griffith had found that there was some information in the bacterial cell which could be transferred from the dead cell to the living cell, changing the (genetic) property of the last one [1]. In 1944 Avery's group found that it was probably the DNA molecule responsible for transmitting the information between these cells [2]. Avery only warily suggested that it could be the DNA but in fact he proved that. Scientific world was not ready to accept the hypothesis that such a dull molecule, composed of only four different subunits can be a carrier of genetic information. Proteins seemed to be much better and obvious candidates for such an important role. There were other very fundamental discoveries which enabled the discovery of the double helix - the Chargaff's rules. Chargaff found that the number of adenine (A) equals the number of thymine (T) while the number of guanine (G) equals the number of cytosine (C) in any DNA preparation. Moreover, Chargaff found that the ratio $[A+T]/[G+C]$ is constant for a species independently of the tissue where the DNA was isolated from, though the ratio could be different for different species [3]. The letters A, T, G and C will be used in the text for connotation the four nucleotides building the DNA.

Now, more than 50 years after the DNA double helix discovery, it is very easy to say what the rationale should be at the bases for creating the model - in fact, only four of them (Fig. 1.1):

(1) the molecule was a double helix;
(2) the phosphate backbone was on the outside, bases on the inside;
(3) the strands were antiparallel;
(4) specific base pairing keeps the two strands together.

April 25th, 1953 Watson and Crick published in "Nature" a paper entitled "A Structure for Deoxyribose Nucleic Acid". It begins with the sentence "We wish to suggest a structure for the salt of deoxyribose nucleic acid (DNA). This structure has novel features which are of considerable biological interest" [4]. They proposed that the DNA molecule is a double helix that resembles a gently twisted ladder (ten rungs per one twist). The rails of the ladder are made of alternating units of phosphate and the pentose (deoxyribose); the rungs are composed of pairs of nitrogen bases: AT and GC. Here we find this Chargaff's rules, called now the deterministic complementarity rules or parity rules I (see Fig. 1.1). The DNA model is very elegant but there is one very peculiar feature of the DNA molecule - resulting from the third assumption - the rails are antiparallel. To be antiparallel means that the rails have their directions. Deoxyribose is an asymmetric molecule with five atoms of carbon (numbered 1' - 5'). Phosphate links the atom number 3' of one deoxyribose with the atom 5' of the other one. It is obvious that at one end there is a free carbon number 5' while at the other end there is a free carbon number 3'. For chemists it is enough to say that the whole DNA strand has a direction 5' to 3'. At the end of the double helix DNA molecule one strand has 3' end and the other one has 5' end. Do you think it is not so important? Imagine, you are in the middle of the molecule and you try to reach the end of it - where is the end? Even if you know that the end is marked by the free 3' carbon - but, of which strand? Thus, it is very important to determine the direction in the double strand or, if we accept the direction of single strand $5' \Rightarrow 3'$ then we have to indicate the strand we are thinking about. Please keep in mind this feature because a lot of phenomena discussed in this chapter will show the consequences of this very nuance in the DNA structure for its coding capacity, control of information expression and evolution.

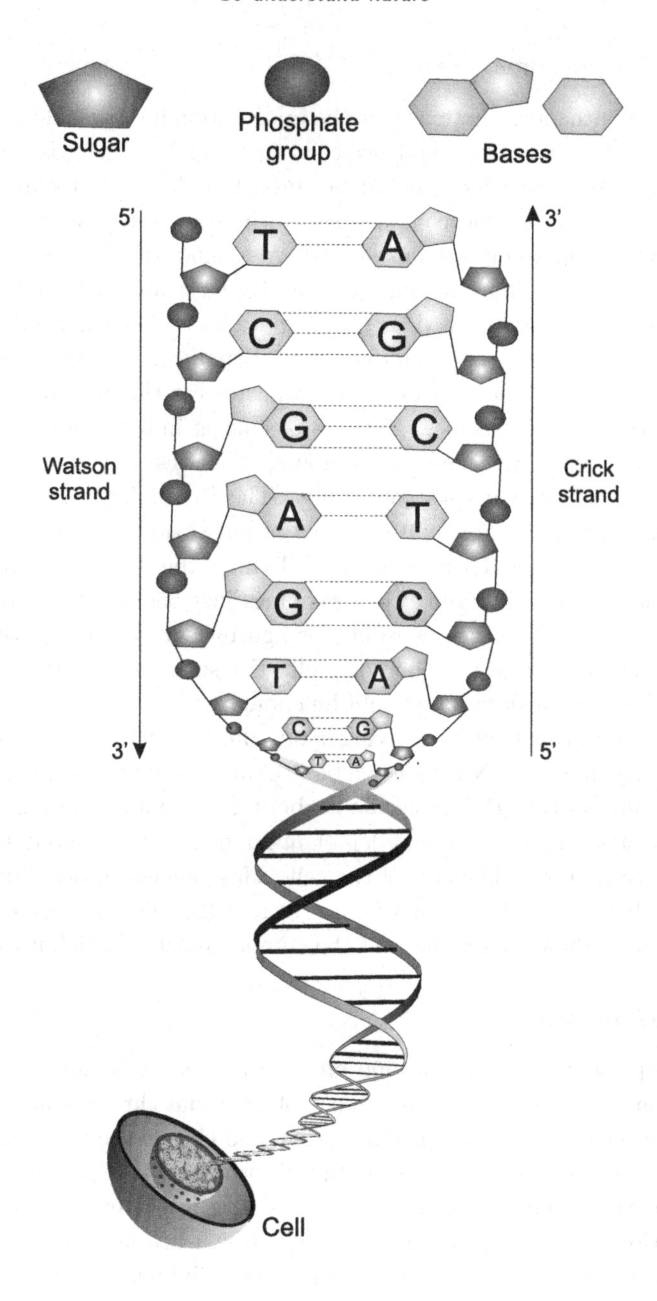

Fig. 1.1. Schematic structure of DNA molecule. Bases: A, T, G and C stand for Adenine, Thymine, Guanine and Cytosine respectively, they form the rungs of the "ladder" while phosphate groups and sugar molecules (deoxyribose) built the rails. Notice two antiparallel strands forming the double helix. The diameter of helix is about 2 nm and length could be measured even in centimeters or several millions of base pairs.

1.2.2. *Chromosomes*

We will see in the next sections how the information imprisoned in the DNA molecule can be expressed and used by the "body" of the host organism. Now, the only issue which should be important for us is to imagine the DNA molecule inside the compartments where we can find it. If we are talking about the living organisms and we assume that viruses don't live (I am not sure if I know why), then the bacteria are the smallest living creatures which use DNA as their genetic data base. The whole information of one bacterial cell is called its genome. Usually it is built of one DNA molecule though, some species of bacteria possess the genome composed of more than one molecule. If the molecule is indispensable for life of the bacteria cell it is called chromosome, otherwise it is called plasmid. We will talk about chromosomes, only. The free living bacteria usually have larger genomes - up to 10 millions of nucleotides (!). Some bacteria have found a very comfortable niche for living - the interior of other cells. Those bacteria usually reduce their genomes just because they don't need the whole information indispensable for fighting for everything; they have almost everything ready in the host cell. For such bacteria 0.5 million of nucleotides in their data base could be enough.

Escherichia coli, bacterium living in our alimentary tract is in the middle of this range and has DNA molecule built of about 5 millions of nucleotides. The diameter of the DNA molecule is about 2 nm and the length ... about million times larger - of the order of millimeters. It is about thousand times more than the diameter of the cell. This range of proportion is very common for bacterial as well as eukaryotic cells, DNA molecule is three orders longer than the diameter of the compartment in which it exists.

1.2.3. *Genomes*

Here we present some information on the bacterial genomes. *Escherichia coli* chromosome is circular, like most of bacterial chromosomes. All information necessary for life is enclosed in the chromosome. Nevertheless, bacteria can possess a lot of additional information indispensable for surviving in the special conditions like antibiotics in the environment. This information can be encoded in plasmids, which are usually also circular and sometimes as large as chromosome. The difference between plasmid and chromosome is in the control of their replication. Replication of chromosomes is both, positively and negatively controlled and perfectly synchronized with the cell division and simply it is impossible to imagine the living cell without chromosome - without plasmid it is possible. Thus, we have to replicate the chromosome before we try to divide the cell.

1.2.4. *Topology of DNA replication*

The topology of DNA replication, even of such a small molecule as *E. coli* chromosome, is a good opportunity for mathematicians to exercise their imagination - usually the abstract thinking satisfies them entirely. Imagine a rope in your living room, 1 cm in diameter and 10 km long, ends tied. The rope is made of two lines wrapped around themselves 500,000 times. During the replication you have to separate the two lines and to connect the new lines with each of the separated one. *E. coli* can do all that in 45 minutes. During this time it is able to build into the DNA 10 millions new nucleotides (five millions into each of the new strands). To make the exercise more demanding - *E. coli* can divide every 20 minutes in the favorable conditions and need 45 minutes to replicate its chromosome. After a few generations its genetic information would be diluted. Do you imagine why it is not a case? Because the replication cycles start at 20 minute intervals, chromosome can replicate in many points simultaneously.

1.2.5. *DNA asymmetry*

The second problem is connected directly with the mechanism of DNA replication. Replication is performed by the specific biochemical machinery with a speed higher than 1000 nucleotides per second. This machinery is able to build the strand only in one direction from 5' to 3' end. Additionally, synthesis of DNA has to start with a primer - a short fragment of another kind of nucleic acid - RNA. The topology of the double helix replication is shown in Fig. 1.2. Double helix has two different ends of strands at each end. Replication starts at the nonrandom point called "Origin of replication" (Ori) and, at the beginning it proceeds only on one old strand starting from its 3' end. Since the synthesized strand is antiparallel - the new strand has 5' end at the start point (it is called "leading strand"). The second old strand stay free - "single stranded" until the replication move about 2000 nucleotides. Then a starter is synthesized in the replication fork and the second new strand starts to synthesize - it is called "lagging strand".

Meantime the replication of the leading strand proceeds for further 1000 - 2000 nucleotides and another starter for the second fragment of lagging strand is synthesized. Next, starters between DNA fragments of lagging

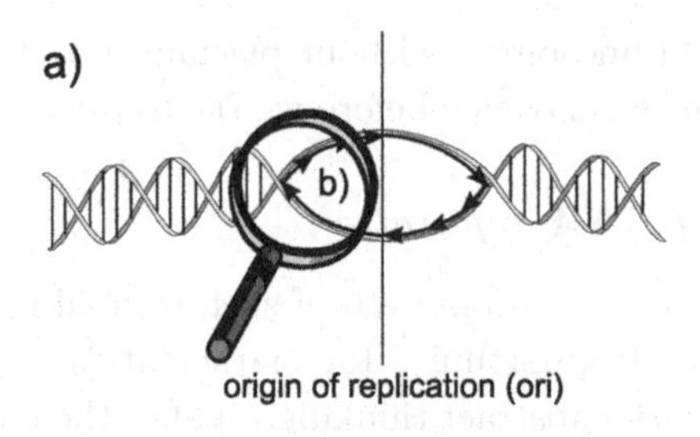

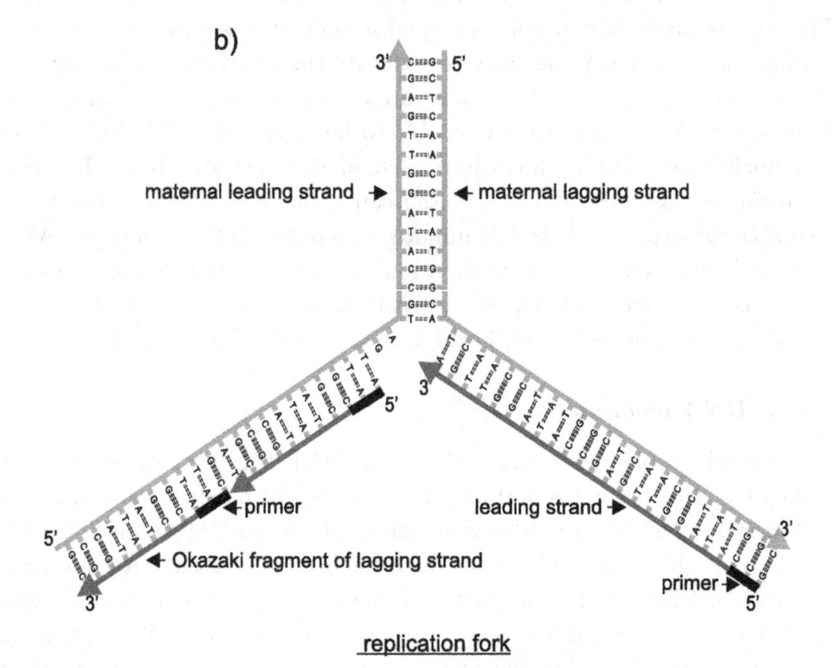

Fig. 1.2. Schematic representation of DNA replication. Panel "a" shows the loop close to origin of replication - oriC where the replication starts in two directions. At this point the two DNA strands have their "switches" from leading to lagging mode of replication. The lower panel shows the differences in mechanisms and topology of replication of the leading and the lagging DNA strands in the region of replication fork. The two replication forks shown in the upper panel will meet at the opposite pole of circular DNA chromosome - terminus of replication.

strands are cut off and gaps are filled up with deoxyribonucleotides. There are in fact two different mechanisms of DNA replication - one for the leading strand and the other one for the lagging strand. Of course, during the process of replication the new nucleotides are built into the new strand according to the rules of complementarity: A-T and G-C. There are many reasons in the whole process of replication for which more or less random errors are introduced during the replication when the rules of complementarity are not fulfilled. These errors - called mutations - are rather rare events, they happen with a frequency of the order of one per genome replication, independently of the genome size. The most frequent errors are substitutions, when the wrong nucleotide is placed in the new strand. For example instead of A opposite to T the G, C or T can be placed. There are three wrong choices for each of four nucleotides. Thus, there are 12 different substitutions. Since the mechanisms of replication of the leading and lagging strands are different, the distributions of the most often substitutions are also different. After many generations, the specific bias of substitutions results in the corresponding bias in the nucleotide composition of the two DNA strands. They are complementary but their nucleotide compositions are different. We call this feature of DNA - "DNA asymmetry".

It is easy to prove that DNA is asymmetric. You remember Chargaff's rules resulting from the complementarity, called sometimes parity rules I: the number of A equals T and the number of G equals C. These rules are deterministic. Let's perform virtual experiment; take a natural DNA sequence, disrupt it for single nucleotides and remake the molecule drawing randomly nucleotides from the whole pool where initially the number A=T and G=C. If we try to build a DNA molecule having such a balanced numbers of nucleotides, using them randomly but keeping the complementarity rules, we can expect that the number of A would equal T and the number of G would equal C in each strand, the differences should be statistically negligible. This stochastic rule is called parity rule II. The natural DNA molecules usually do not fulfill the parity rule II - they are asymmetric. The simplest way to visualize the DNA asymmetry is to perform the DNA walk (Fig. 1.3).

We can put a walker on the first position of the DNA strand and declare that it is moving in the two-dimensional space according to the rules: A [1,1], T[1,-1], G[1,0], C[1,0]. In such a walk, called A/T walk, the walker is going up if there are more A than T in the strand. For the G/C walk the corresponding moves of the DNA walker are: A [1,0], T[1,0], G[1,1], C[1,-1] and walker is going up if there are more G than C in the DNA strand.

Two examples of the DNA analysis using the DNA walks are presented in Fig. 1.4. The walk shown in panel b was done for random DNA sequence of the global nucleotide composition characteristic for real *E. coli* chromosome but randomly distributed along the two strands obeying only

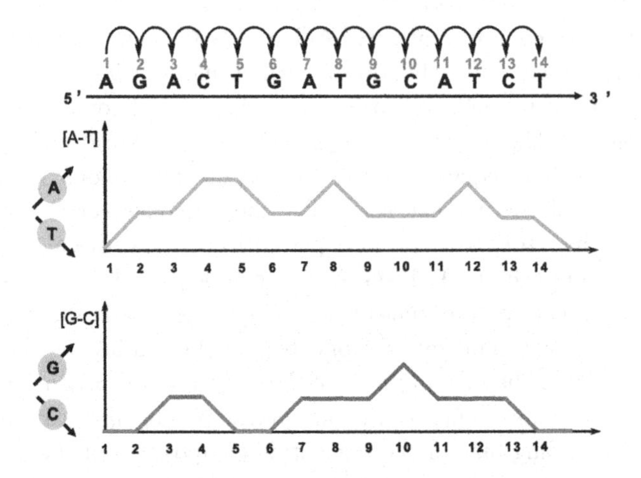

Fig. 1.3. A-T and G-C DNA walks. In the A-T type of a walk, walker starts from the first nucleotide of the analyzed sequence and its moves are: A [1,1], T[1,-1], G[1,0], C[1,0]; in the G-C type of walk its moves are: A [1,0], T[1,0], G[1,1], C[1,-1]. DNA walks of this type show local and global asymmetry of DNA molecule.

the parity rule I (complementarity). This is a sequence obtained in our "virtual experiment" described above. The second walk was done for the real *E. coli* chromosome. The "real *E. coli* chromosome" means a sequence of nucleotides obtained during the experimental sequencing of *E. coli* chromosome, (available in the genomic data bases). Note that a sequence of nucleotides of one strand is different than the sequence of the other one but the first strand determines the sequence of the other one unambiguously. Thus, it is enough to put the sequence of only one strand into the data bases. This strand is named the Watson strand, the other one (complementary) is called the Crick strand. The beginning of the circular chromosome is chosen arbitrarily and it is presented in the direction $5' \Rightarrow 3'$. It is just for better communication when describing some additional features or sequences on the chromosomes. Comparing the last two DNA walks (randomly arranged sequences and the real one) it is clear that that *E. coli* chromosome does not obey the parity rule II. It is asymmetric [5]. It is divided by Ori and Ter regions for two parts, called replichores. At the points Ori and Ter the asymmetry changes its sign. Again, comparing the topology of replication with the results of DNA walks we can conclude that at the Ori the DNA strands change their role in replication from the leading to the lagging and vice versa. Concluding; two strands of one double helix DNA molecule are under different mutation pressure and have different nucleotide composia)

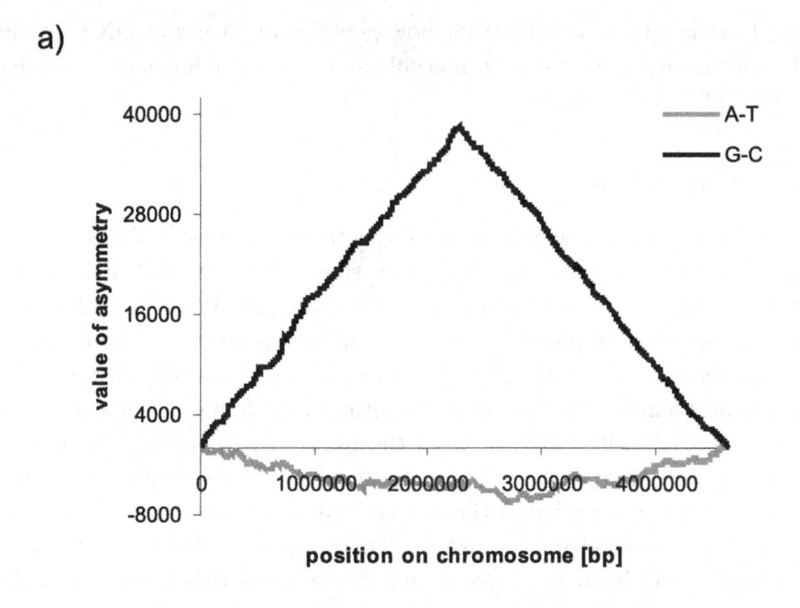

b)

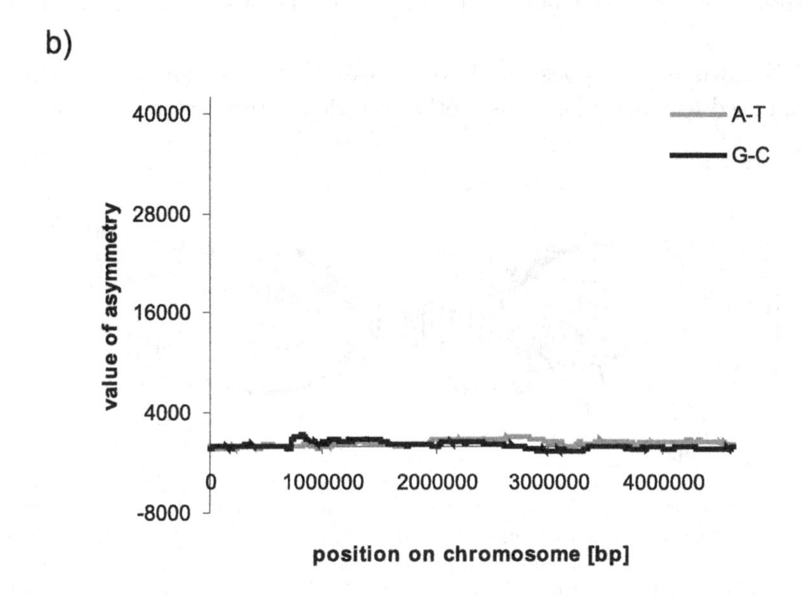

Fig. 1.4. Asymmetry of *Escherichia coli* DNA chromosome; (a) - A-T and G-C DNA walks performed for the real *E. coli* chromosome about 4.5 million base pairs long. Walkers start at the terminus of replication, the extremes (very well seen maximum for G-C walk) are in the origin of replication region; (b) - DNA walks performed for the random DNA sequence of the same general nucleotide composition as *E. coli* chromosome. No significant DNA asymmetry is seen in the same scale.

tion. That is why it is important how genes - fragments of DNA double helix - are positioned on such a molecule and how the information encoded into the DNA is deciphered.

1.2.6. *Transcription*

The DNA molecule is a data base of genetic information. To use it, the proper information from this data base should be retrieved, copied and translated into the function. Usually, "the function" is performed by the protein molecule composed of many subunits (up to many thousands) - amino acids. There are 20 different amino acids built into the proteins. We omit the problem of how "the proper" information is found on the huge DNA molecule and we will start just from the point of its copying. To express the desired information encoded in the sequence of deoxyribonucleotides, the very sequence is copied in the process called transcription into another nucleic acid - ribonucleic acid (RNA). There are some differences between DNA and RNA; RNA is single stranded, possesses ribose instead of deoxyribose and Uracil (U) instead of T, complementing with A. RNA is also synthesized from 5' to 3' end on the antiparallel DNA strand (see Fig. 1.5).

RNA sequence suppose to have "sense" thus, the antiparallel DNA strand used as matrix for transcription is called antisense strand while the

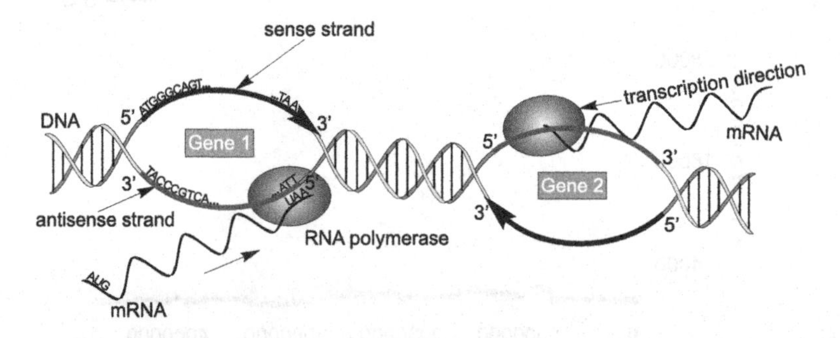

Fig. 1.5. Transcription process. RNA polymerase transcribes the antisense DNA strand. The transcript - RNA sequence - corresponds to the sense DNA strand. Notice that the direction of RNA strand (5'- 3') is the same as the direction of the DNA sense strand. If in the left transcription bubble the sense DNA strand is a leading strand - it is said that the genes is located on the leading strand but notice that the gene transcribed in the left bubble is located on the lagging strand and the direction of its transcription would be different that the direction of replication fork movement. This topology is very important for many mechanisms of gene expression.

other one (complementary) is again called a sense strand. The position of sense strand of protein coding sequence and its direction describe the topological parameters of gene on chromosome. If the sense strand of gene is located on the leading strand, its transcription has the same direction as the replication fork movement and it is said that "the gene" is located on the leading strand. If the sense strand is located on the lagging strand then the direction of transcription is opposite to the replication fork movement. The RNA sequence is translated into amino acid sequence of proteins. The basis of the complicated decoding system is the genetic code.

1.2.7. *Genetic code, degeneracy, redundancy*

Nucleic acid is composed of four different subunits while proteins are built of 20 different amino acids. That is why the genetic code has to use at least three nucleotides for coding one amino acid (and really it uses tri-nucleotides) resulting in 64 possible different triplets - much more than the minimum number. The tri-nucleotide coding one amino acid is called "codon" (or triplet). The meaning of all codons is shown in the table of genetic code (Fig. 1.6).

There are many peculiar properties of the genetic code. First of all the genetic code is universal which means that all organisms in biosphere use the same code. Only some of them, usually with very small genomes, use some "dialects" differing in the sense of single codons. We can state that all organisms communicate in the same language.

The other properties of genetic code are: not overlapping, without any commas, with the fixed start and stop of translation, unambiguous and degenerate. For our purposes the last two features of the genetic code are very important. Unambiguous means that any codon codes for only one amino acid while degeneration means that the same amino acid can be coded by more than one codon. One of the hypotheses of genetic code evolution assumes that initially the codon assignments varied and it was the selection pressure which optimized the code in respect to reduction of the harmful effects of mutations occurring during replication and transcription and to minimization the frequency of errors during translation process [6, 7] for review see [8]. During further evolution connected with the increase of genomes, the genetic code was frozen [9] and it was not possible to re-interpret the meaning of codons because every change would affect many amino acid positions in proteins and it would have catastrophic consequences for the organisms. As a result of such an evolution, not only

	U	C	A	G
U	UUU Phe UUC Phe UUA Leu UUG Leu	UCU Ser UCC Ser UCA Ser UCG Ser	UAU Tyr UAC Tyr **UAA Stop** **UAG Stop**	UGU Cys UGC Cys **UGA Stop** UGG Trp
C	CUU Leu CUC Leu CUA Leu CUG Leu	CCU Pro CCC Pro CCA Pro CCG Pro	CAU His CAC His CAA Gln CAG Gln	CGU Arg CGC Arg CGA Arg CGG Arg
A	AUU Ile AUC Ile AUA Ile **AUG Met**	ACU Thr ACC Thr ACA Thr ACG Thr	AAU Asn AAC Asn AAA Lys AAG Lys	AGU Ser AGC Ser AGA Arg AGG Arg
G	GUU Val GUC Val GUA Val GUG Val	GCU Ala GCC Ala GCA Ala GCG Ala	GAU Asp GAC Asp GAA Glu GAG Glu	GGU Gly GGC Gly GGA Gly GGG Gly

Fig. 1.6. **Genetic code.** Notice the way of construction the genetic code table; Codons are grouped in boxes - four codons in each box. Four boxes in one column have the same second nucleotide in codons. Four boxes in one raw have the same first nucleotide in codons. Thus, four codons in one box differ only in the third codon positions. All four codons in each of grey boxes code for the same amino acid (three letters abbreviations of the amino acid names are at the right side of codons). In all other cases pyrimidines in one box (C or U) code for the same amino acid and in only two cases different purines (A or G) change the meaning of codons.

the degeneration of the genetic code is important but also the way it is degenerated.

One mechanism of optimization results directly from the simple structural relations between nucleotides in the double helix - one large and one small nucleotide fit better to form a pair (a rung in a ladder) and it is much easier to overlook the mutation changing a small nucleotide for another small ($T \leftrightarrow C$) or large one for another large ($A \leftrightarrow G$). These mutations are called transitions. The other kind of mutations, when small nucleotides are replaced by large ones or vice versa (transversions), is much rarer. Note that transitions in the third positions of codons change the sense of codons only in two cases: met/ile and trp/opal. All other transitions in the third positions are synonymous and even half of transversions in the third positions are synonymous. Generally, mutations in the third codon positions are usually accepted by selection. Furthermore, mutations in the first position could be also accepted because they change one amino acid for another one with similar chemical properties (i.e. polarity, hydrophobicity). The most deleterious are mutations in the second positions of codons. Such observations suggest that positions in codons vary in both, their role in coding as well as in the evolution rate, resulting in their nucleotide composition.

1.2.8. *Topology of coding sequences*

To analyze the nucleotide composition of coding sequences, the DNA walks can be used. In this version of two-dimensional walks [10], the walkers moves are for: G - [0,1], A - [1,0], C - [0,-1] and T - [-1,0]. The best results are obtained when the walks are generated separately for each position in codons (Fig. 1.7).

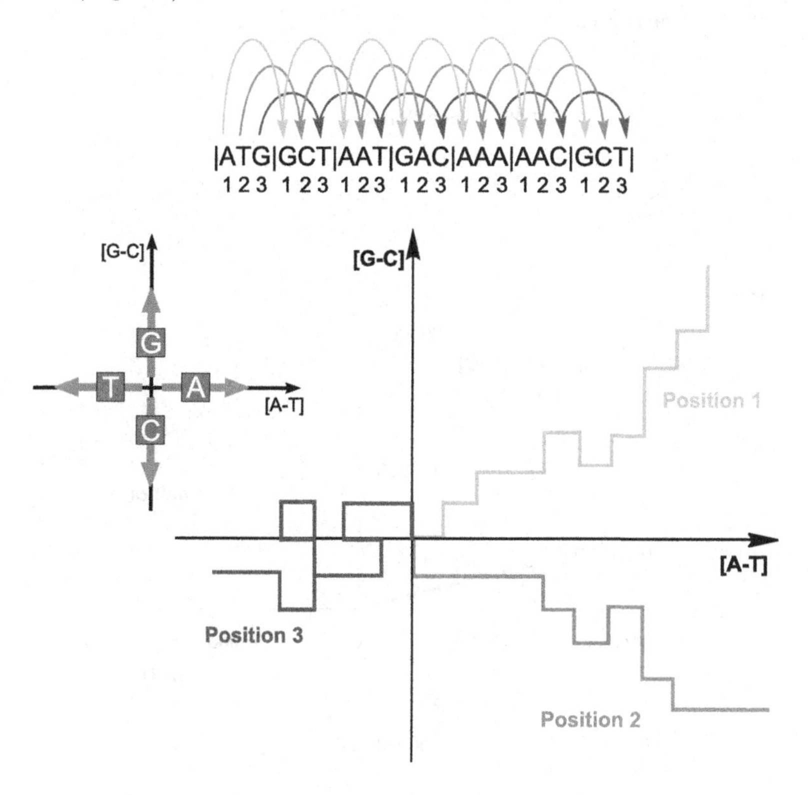

Fig. 1.7. DNA walks on protein coding sequences - spiders. There are three walks in the two-dimensional space. The first walker starts from the first nucleotide, the second walker from the second nucleotide and the third walker starts from the third nucleotide of the first codon. Their jumps every three nucleotides on sequence correspond to moves accordingly to the nucleotide just visited: G - [0,1], A - [1,0], C - [0,-1] and T - [-1,0]. Each walker produces distinct walk called a spider leg numbered according to the position of nucleotides in analyzed codons.

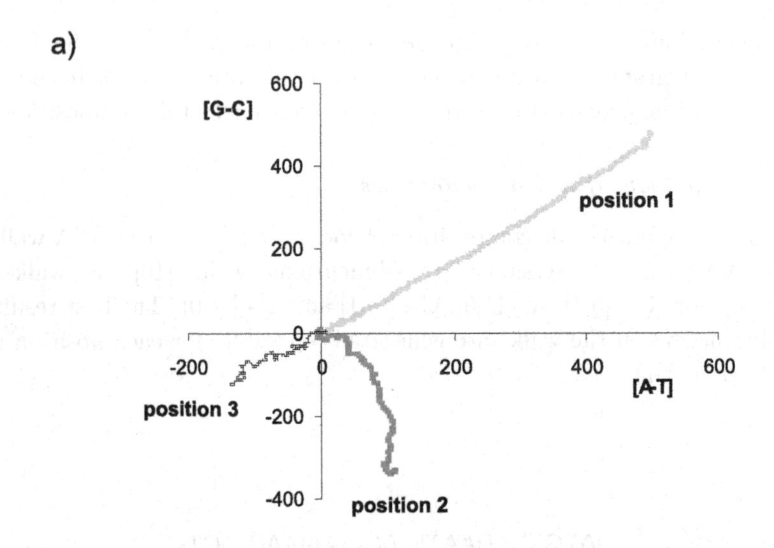

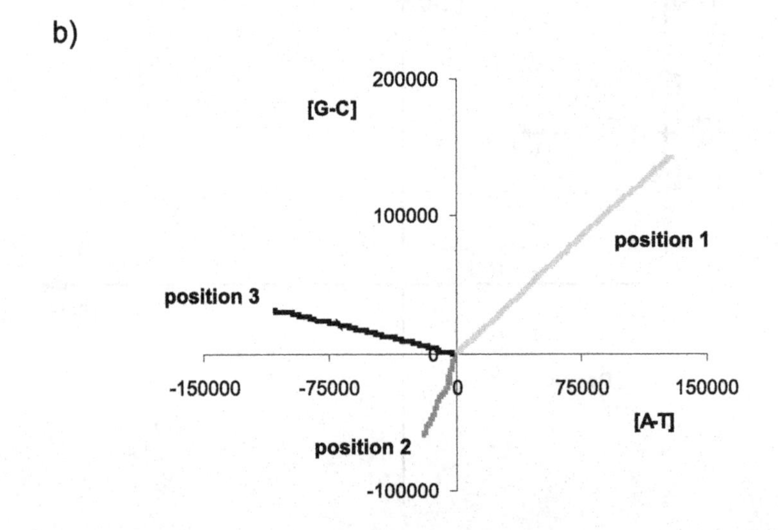

Fig. 1.8. Spiders for a single gene (a) and for all protein coding genes of *E. coli* located on the leading DNA strand spliced together (b). Straight lines in the (b) panel suggest that the trends in nucleotide composition of all three positions in codons are universal at least for the genes located on this DNA strand.

To construct the DNA walk for the first codon positions the walker starts from the first position of the first codon and jumps every three nucleotides until it reaches the first position of the stop codon. For the second codon position the walker starts from the second position of the first codon and for the third positions it starts from the third position of the first codon. Three plots describing a coding sequence are called a spider [11]. The first example of spider shown in Fig. 1.8 represents three DNA walks done for a single gene of *E. coli*. It is clear that the walks are different and correspond to different nucleotide composition of the three positions in codons.

The second spider in Fig. 1.8 shows walks constructed for all genes located on the leading strand of the *E. coli* genome spliced into one sequence. Now, it is seen that trends for one gene are universal for the whole genome though, in larger scale. More about the analysis of coding capacities of genomes can be found at the web site: `http://www.smorfland.uni.wroc.pl/www/dnawalk.html`.

Such kinds of walks can be done also for analysis of distribution of specific codons [12], codons for specific amino acids or groups of amino acids [13] or any other specific sequences occurring along the chromosomes, like sequences controlling or involved in the replication processes [14].

1.2.9. *Mutational pressure*

The DNA walks done for coding sequences show asymmetry of the coding sequences - a very specific asymmetry - different for each position in codons. It is obvious that this asymmetry is generated by selection which forces the specific amino acid composition of functional proteins and specific codon usage - if one amino acid is coded by more than one codon, some preferences of "codon usage" are observed in the genome. A few sections above we have discussed the problem of DNA asymmetry of replichores generated by different mutation pressures for leading and lagging strands. Now it is time to understand what could be the consequence of specific location of coding sequence on chromosome - sense strand of gene located on the leading strand or on the lagging strand. Imagine one example - leading strand is poor in C because this nucleotide is replaced by T preferentially on this strand. On the other hand, the second position of codons is rich in C and this position is very carefully watched by selection. If we put a coding sequence with sense strand reach in C on the leading strand, then it will be very vulnerable for mutations because the probability of substitution C by T is higher and such a substitution could be deleterious for the function of coded protein.

1.3. Evolution of coding sequences

Coding sequences are under mutation pressure which depends on their location on chromosome and under selection pressure which demands the determined function of protein. Analyzing any genome we see the results of compromise between these two forces. To characterize one force we have to describe the other one. We have succeeded in describing the mutation pressure for the genome of *B. burgdorferi* [15–17]. In fact there are two, some kind of mirror, matrices describing the relative frequencies of nucleotide substitutions - one for the leading strand and the other one for the lagging strand (Fig. 1.9).

The matrices are constructed in such a way that at each line in the first column is a nucleotide which, if chosen for substitution, can be replaced by one of three other nucleotides with a given probability, if it is not substituted it stays unchanged. The sum of all 12 substitution probabilities in the matrix equals one.

Thus, the frequency of substitution in a given sequence depends on the probability of choosing the nucleotide for substitution (parameter p_{mut} - independent of nucleotide composition of the sequence) and on the nucleotide composition of the sequence under the mutation pressure. For mutation pressure characteristic for leading strand described by Fig. 1.9a the highest possible frequency of substitution would be obtained for a sequence composed of C only. If there is no selection pressure, means any constraint set on the sequence under mutation pressure - the sequence would reach the equilibrium composition corresponding to the mutation pressure.

In Fig. 1.10 we show the dynamics of reaching this state starting from random DNA sequence with the uniform nucleotide composition. The mutational pressure was for leading strand of *B. burgdorferi* genome. The final nucleotide composition of the DNA sequence corresponds to the nucleotide composition of the third codon position in the coding sequences located on the leading strand of this genome. One can conclude that the third coding positions in this very genome are not under the selection pressure - they are freely shaped by the mutational pressure. The difference (distance) between the nucleotide composition of such sequences, without selection pressure, and any other sequences in the real genome could be considered as a measure of selection force.

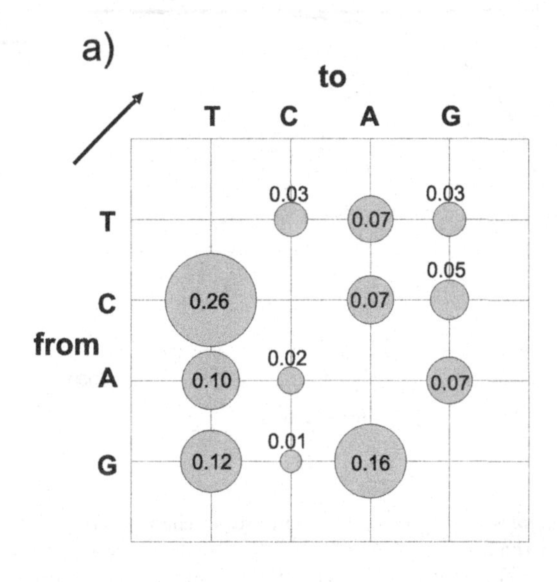

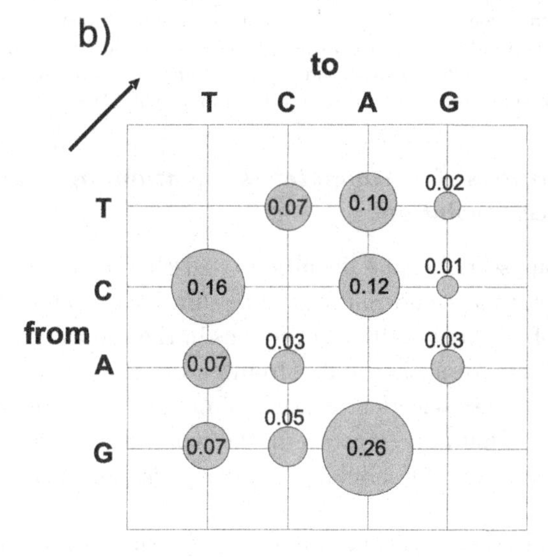

Fig. 1.9. Substitution matrices for the leading (a) and the lagging (b) DNA strands of the Borrelia burgdorferi chromosome. Mutation can replace any nucleotide from the left column by one of the three other nucleotides (upper raw) nevertheless, the substitution probability is different for both, substituted nucleotide (sum of the raw) and substituting nucleotide (numbers in columns). Data in the tables show the relative substitution rate - sum of all numbers equals 1. The real substitution rate in the genome is of the order of million times lower.

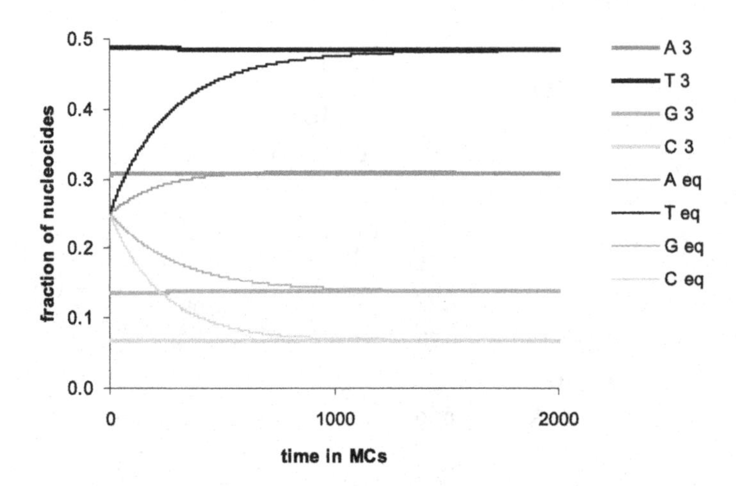

Fig. 1.10. Dynamics of the nucleotide composition changes under mutational pressure characteristic for the leading DNA strand of *B. burgdorferi* chromosome. There are two different starting compositions of DNA sequences; the first one corresponds to the third position of codons (does not change in the course of simulations) and the second one corresponds to the random sequence with equimolar nucleotide composition (fraction of each nucleotide equals 0.25). The second type of sequence has reached the composition characteristic for the third codon position after about 1000 MCs.

1.3.1. *Conditions for computer simulation of coding sequences evolution*

In the previous sections, the definitions and characteristics of coding sequences, genetic code and mutation pressure have been described. The composition of all three positions of codons in the coding sequences usually are different, moreover they differ from those generated by pure mutation pressure because selection pressure eliminates some configurations. To check the effect of mutation pressure on the evolution of coding sequences we have performed computer simulations under following conditions [18, 19]:

(1) Mutational pressure has been described by substitution matrix for the proper DNA strand, if the coding sequence was located on the leading strand, the used substitution matrix was for the leading strand, if coding sequence was located on lagging strand the substitution matrix was also for lagging strand (1.9 2).
(2) The original coding sequence was translated into amino acid sequence

and fraction of each amino acid has been counted.

(3) In one Monte-Carlo step, each nucleotide of the gene sequence was drawn with a probability p_{mut}, then substituted by another nucleotide with the probability described by the corresponding parameter in the substitution matrix.

(4) After each round of mutations, the nucleotide sequences were translated into amino acid sequences and compared to the product of the original gene. Selection pressure was introduced as tolerance for amino acid substitutions. For each gene the selection parameter (T) was calculated. T is the sum of absolute values of differences in fractions of each amino acid between the original sequence (f_0) and the sequence after mutations (f_t): $T = \sum_{i=1}^{20} |f_0^i - f_t^i|$. It describes deviation in the global amino-acid composition of a protein coded by a given gene after mutations, in comparison to its original sequence from the real genome. If T was below an assumed threshold, the gene stayed mutated and went to the next MC step, if not, the gene was "killed", which means eliminated from the genome and replaced by its allele from the second genomic sequence, originally identical, simulated in the part of the same MC step.

(5) The number of all substitutions which occurred during the simulation, the number of accumulated substitutions (accepted), and the number of replacements of genes (killed genes) were counted after each MC step (Fig. 1.11).

The maximum tolerance used in these simulations corresponded to the mean value of divergence between orthologous genes from *B. burgdorferi* and *Treponema pallidum* - a related to *Borrelia*, completely sequenced bacterium [20]. Orthologous genes mean genes which had a common ancestor sequence in the past. Notice that the mutation process in the model is separated from selection by translating the nucleotide sequences into amino acid sequences thus, the important role in the whole evolutionary process is played by the genetic code.

1.3.2. *Dynamics of mutation accumulation and gene's elimination*

At start of simulations all sequences have the original amino acid composition and T values equals 0. In the course of simulations, mutations accumulate into the nucleotide sequences and, when translated into amino acid

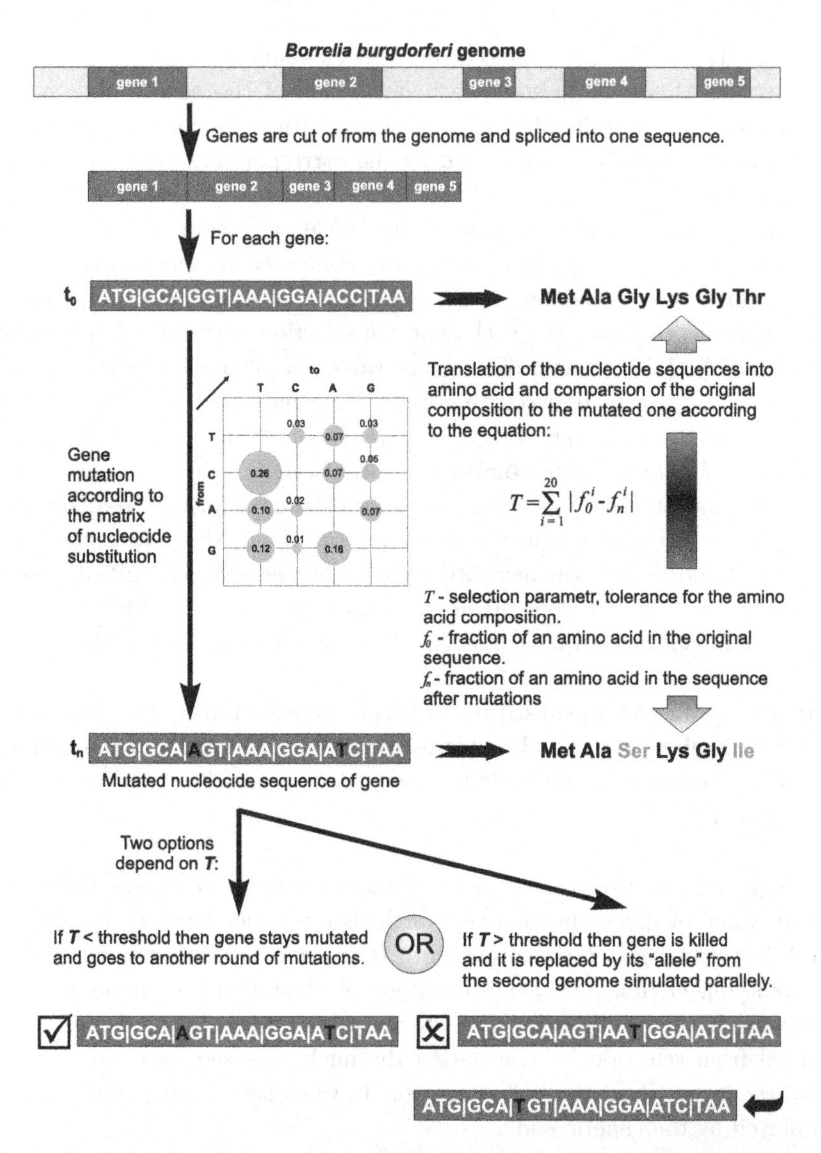

Fig. 1.11. Simulation of the evolution of protein coding sequences. All genes from the B. burgdorferi genomes were cut off the genome and spliced into one sequence. Each gene sequence was translated into amino acid sequence. During each MC step nucleotides were drawn with probability p_{mut} and substituted according to the proper substitution matrix (for leading or lagging strand) and again gene sequences were translated into amino acid sequences. For each of the two amino acid sequences the fractions of amino acid were counted and compared. If the sum of differences was higher than the assumed tolerance - the gene was eliminated and replaced by its homolog from parallel evolving genome otherwise it passed to the next round of simulation.

sequences, they eventually trespass the assumed tolerance. The nucleotide sequence is then eliminated from the pool and replaced by its orthologous sequence from the parallel evolving sequences. Thus, at the beginning the elimination rate is relatively low and it is growing with time of simulation. This is negatively correlated with the probability of acceptance the mutation into the coding sequence - at the beginning the mutation accumulation rate is relatively high and it is decreasing with time of simulation (Fig. 1.12).

This seems to be obvious, but it is some kind of artifact because it has been assumed that at the beginning of simulations all sequences are equally distant from their tolerance limit. Nevertheless, the sequences diverge from the original ones during the evolutions accumulating mutations which are not totally random - mutations are introduced accordingly to the substitution matrix. In Nature, it is a known phenomenon of inverting the DNA sequences. If a coding sequence is inverted at the same position (without translocation to another replichora) it is said that its location is changed from the leading strand to the lagging strand or *vice versa* (see Sec. 1.2.9). Such inversion results in replacing the mutation pressure characteristic for the leading strand to that of lagging strand (or *vice versa*) [21, 22]. We have tested the effect of such translocations. When the mutation pressure was changed every MC step, the elimination rate of sequences dropped when compared with stable mutation pressure [23]. Moreover, it was associated with the increased accumulation of mutations. The same results were obtained for sequences originally located on the lagging strand.

Nevertheless, in Nature, the frequency of inversions is rather low and it corresponds to about 200 MCs in simulations under our p_{mut} value [24]. We have checked the evolution rate of sequences under such a changing regime of mutation pressure. The results of simulations for sequences originally located on the leading strand are shown in Fig. 1.12. Notice that inversions help to escape coding sequences from elimination by selection. The effect is very strong and it is again associated with higher accumulation of mutations in the coding sequences.

Results of simulations suggest that genomes could use this strategy to lower the costs of mutation pressure and to increase the biodiversity (divergence, mutation accumulations).

To look for such an effect in Nature we have analyzed the orthologous sequences of some pairs of closely related genomes. We have compared the divergence of corresponding genes in a pair of genomes grouped into classes: in both genomes orthologs stay on the leading DNA strand or, on

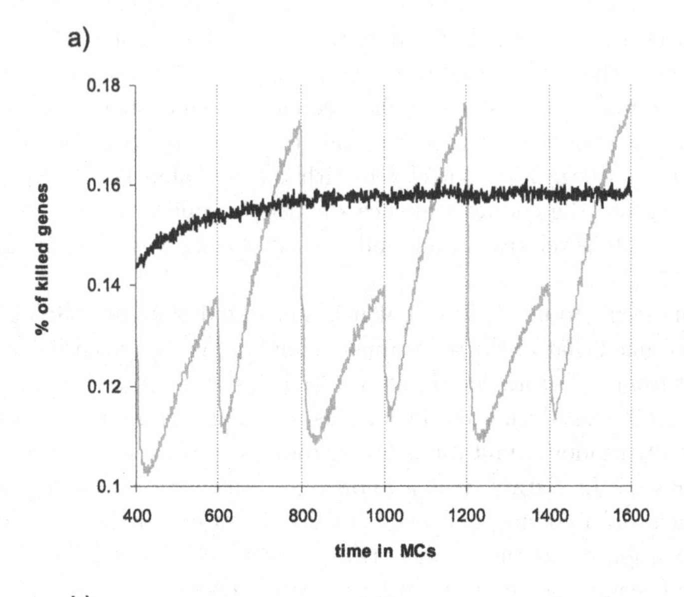

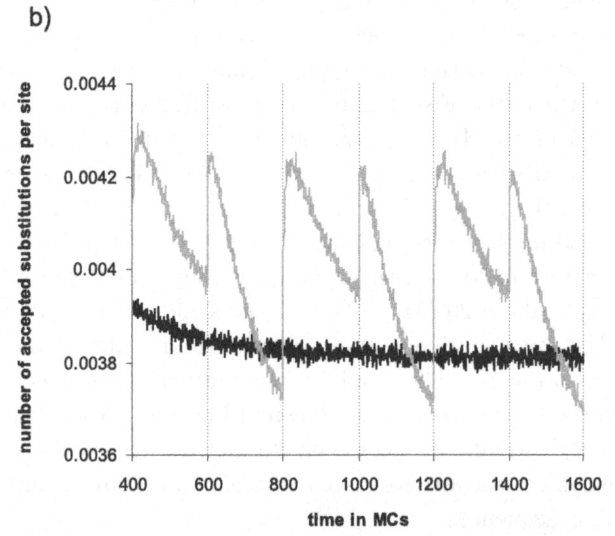

Fig. 1.12. The effect of inversions on the evolution of coding sequences. Panel (a) the killing effect of mutational pressure; solid, black line without inversions, gray line - with inversions every 200 MCs. Panel (b) accumulation rate of amino acid substitutions. Notice that each inversion causes drastic decrease of deleterious effect of mutations (killing effect) and simultaneous increase of accepted amino acid substitutions. Generally, inversions decrease the deleterious effect of mutations and increase the evolution rate of coding sequences measured by the amino acid substitution rate (divergence of protein sequences).

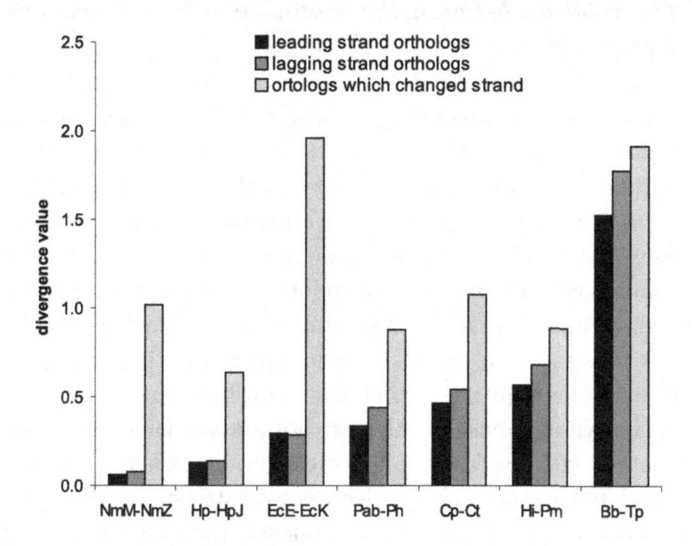

Fig. 1.13. Divergence rate of protein sequences coded by genes located on the leading strands, lagging strands or genes which changed their locations since the speciation. The first three blocks of columns show the divergence of orthologous sequences in very closely related species or even strains of the same species of bacteria. The difference between the divergence rates of inverted orthologs is much higher than those which haven't change their positions. Abbreviations: *Escherichia coli* K12-MG1655 (EcK) - *E. coli* O157:H7 EDL933 (EcE); *Helicobacter pylori* 26695 (Hp) - *H. pylori* J99 (HpJ); *Neisseria meningitidis* MC58 (NmM) - *N. meningitidis* Z2491 (NmZ); *Chlamydia pneumoniae* (Cp) - *C. trachomatis* (Ct); *Pyrococcus abyssi* (Pab) - *P. horikoshii* (Ph); *Borrelia burgdorferi* (Bb) - *Treponema pallidum* (Tp); *Haemophilus influenzae* (Hi) - *Pasteurella multocida* (Pm);

the lagging DNA strand or, they changed the location - in one genome a sequence is located on the leading strand while in the other one on the lagging strand [25]. At very short distances (closely related bacteria strains), we can assume that only one translocation happened during the time of separated evolution. The results of genomic analysis are shown in Fig. 1.13.

They are in agreement with the results of simulations. Sequences which were inverted since the divergence of species (or strains) have the highest relative divergence - they were not eliminated and they accumulated more mutations. Sequences which have not been inverted accumulated fewer mutations because they were eliminated by selection more frequently. At larger evolutionary distances this effect is weaker because in each class of sequences we can expect genes which were inverted several times.

1.3.3. *The relation between the mutation rate and sequence divergence*

It seems that sequence divergence should be highly correlated with the mutation rate. In fact it is not true. It depends on the tolerance of the sequence. If the limit of tolerance is very small, even single mutation can eliminate the sequence and the mutation accumulation will not be observed. On the other hand, in our simulation we have assumed that all mutations which do not change the meaning of codon are accepted, they are neutral. Thus, it seems that at least all these synonymous mutations should be accepted, but it is not true, neither. If mutation rate is too high it could happen that in one sequence two or more substitutions occurred, some of them could be synonymous but other are not tolerated and eliminate the sequence together with the "acceptable" substitutions. There is another effect in the mutation accumulation; if selection tolerates substitution of a given fraction of amino acids in the protein sequence, then the shorter sequences should be more vulnerable to mutations because in long sequences there is still some possibility that mutation in one place could be suppressed by the mutation which occurred at the other place in the "inverse direction". Genomic analysis of the divergence rate of orthologous sequences have shown that the mutation accumulation rate does not depend on the coding sequence length.

Analyzing the vulnerability of coding sequences on nucleotide substitutions it is necessary to consider the generation of stop codons inside the protein coding genes and elimination of start codons. Both kinds of mutations shorten the coding sequences and it is legitimate to assume that they are deleterious and eliminate the gene. The probability of elimination of start codons is independent of the length of coding sequences because each coding sequence possesses only one start codon, while generation of stop codons inside coding sequences linearly depends on the number of codons. These specific properties and effects of substitutions allow estimating some parameters of mutational pressure. Longer sequences are more tolerant for amino acid substitution than shorter ones. On the other hand, the shorter sequences are less vulnerable to the stop codon generation inside them. Thus, it should be possible to find such a set of parameters of mutation rate and tolerance for amino acid substitution where the divergence rate of coding sequences during simulation would not depend on the gene length - just like in natural genomes. In general, the mutation rate should be lower than 1 per selected sequence and tolerance roughly corresponds to that one

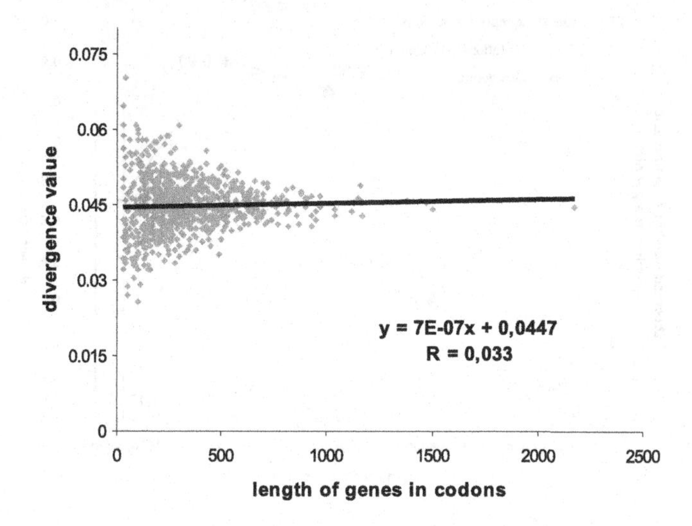

Fig. 1.14. The relation between the divergence rate of coding sequences and their lengths. In Nature there are no significant difference between the length of coding sequences and their divergence rate. To get such an effect in simulations, mutational pressure was lower than 1 mutation per coding sequence in 1 MCs, negative selection for generation the stop codons inside the sequence and elimination of start codons was established, tolerance for amino acid composition was 0.3.

we have counted on the basis of divergence between *B. burgdorferi* and *T. pallidum* - closely related species of bacteria (Fig. 1.14).

1.4. Evolution of whole genomes

In the previous section, the evolution of single coding sequences was described. Now we would like to describe simulations of evolution of the whole genomes. In particular we would like to answer the questions how the divergence rate depends on the mutation rate and the number of co-evolving genomes. We have used the same method of simulations like in case of single coding sequences evolution. The only difference is in the effects of deleterious mutations. In the new approach, mutation which eliminates a single coding sequence eliminates the whole genome. The results of simulations showing the effect of different mutational pressure on the divergence rate, accumulation of mutation and elimination of genomes are shown in Fig. 1.15.

Divergence is counted by alignments the original amino acid sequence

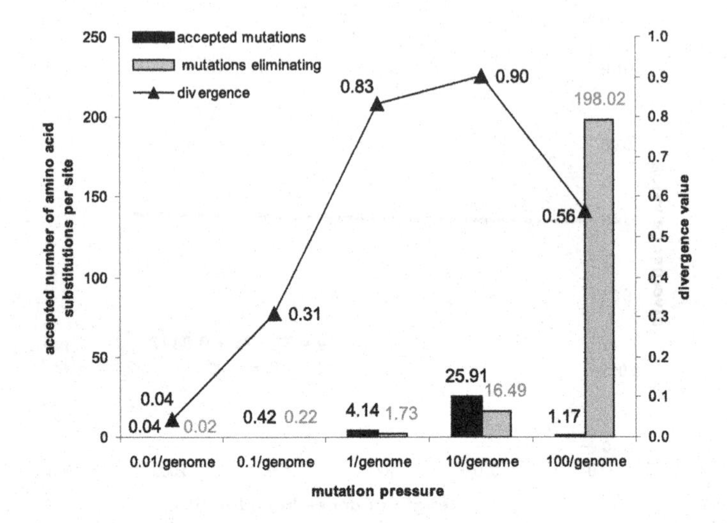

Fig. 1.15. The relations between the mutational pressure and different parameters describing the evolution rate of coding sequences. Notice that the left y-axis shows the total number of accepted (or eliminating) mutations which happened during the simulations. The left scale of y-axis shows the fraction of positions which are different in the coding sequences before and after the simulations (divergence).

with a sequence obtained after mutations; it is the fraction of amino acid positions occupied by different amino acids in the compared protein sequences. Notice that even if two different sequences are aligned some positions can be randomly occupied by the same amino acid thus, divergence cannot be higher than 1.0. Results shown in Fig. 1.15 represent the number of accepted amino acid substitutions which have occurred during the whole simulation - many multiple substitutions have been observed; even reversions were possible - amino acids which have been substituted by other ones have been reintroduced by further mutations at the same position. The number of accepted mutations grows almost linearly with mutational pressure up to 1 mutations in 1 MCs per genome, the acceptance slow down for mutational pressure of the order of 10 per genome per MCs and dramatically decrease with higher mutational pressure.

If the mutational pressure was higher, the number of accidents of elimination the genome by deleterious mutations was also higher and the number of accepted mutations decreased, even if these mutations are neutral. If we accept, as a measure of the evolution costs, the ratio between the number of accepted mutation and the number of mutation eliminating the genome, the optimum of mutational pressure was 1 per genome per generation (Fig. 1.16, Fig. 1.17).

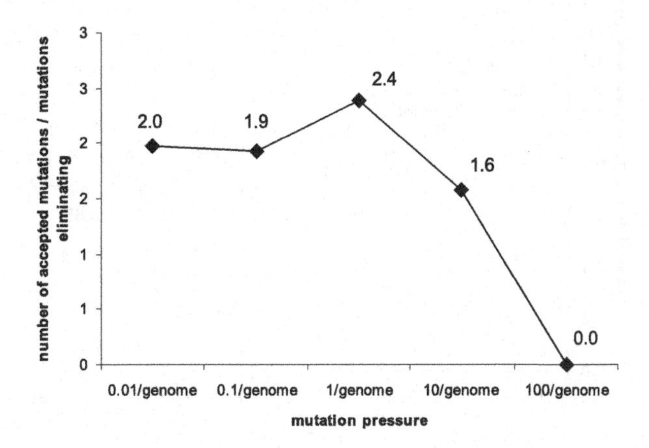

Fig. 1.16. The effect of mutational pressure on the ratio between the number of accepted mutations and eliminating mutations.

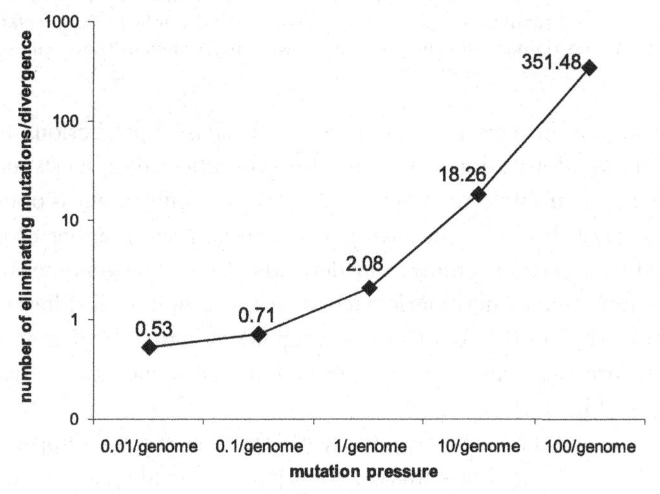

Fig. 1.17. The effect of mutational pressure on the evolutionary costs. It has been assumed that the divergence is a measure of evolution rate and evolutionary costs could be estimated by the ratio between the number of eliminating mutations and divergence. Compare with the results presented in Fig. 1.16.

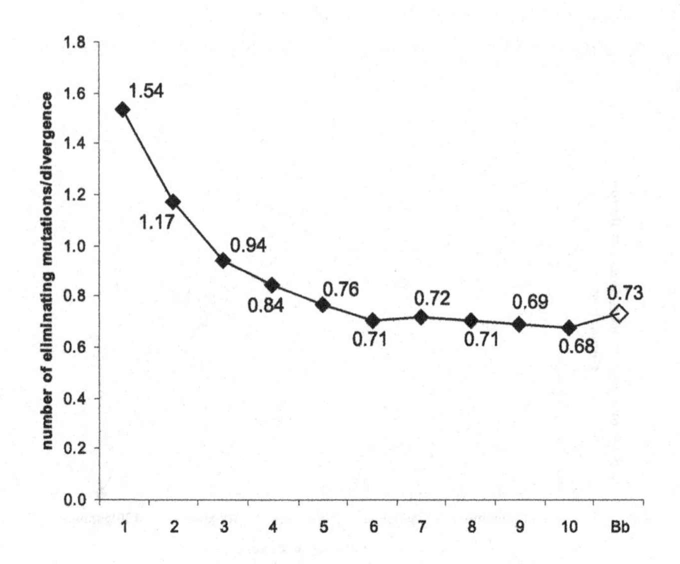

Fig. 1.18. The relation between the evolutionary costs and the genome size. The mutational pressure was constant (1 mutation per genome per MCs). The smallest "genome" (the first from the left - 1) was represented by integer number (85) of *B. burgdorferi* genes and it was approximately 10 times smaller than the whole *B. burgdorferi* genome (the last one at the right). All other "genomes" were multimers of the smallest one.

This value is in agreement with some physicists' predictions based on the stability of information as well as with experimental data estimating the mutational rate in different genomes [27–29]. The effectiveness of evolution counted as ratio between the divergence rate and the number of genomes eliminated by deleterious mutations depends also on the genome size. Next series of simulations where performed with the genomes of different sizes - they increased from 0.1 to 1.0 of the original *B. burgdorferi* genome. The mutational pressure was the same per genome per generation. Results are presented in Fig. 1.18.

The evolutionary costs dropped with the genome size approximately twice. If the mutational rate was constant per nucleotide per generation, the evolutionary costs grow with the genome size (not shown). In conclusion, the increase of the genome's size has to be accompanied with the higher accuracy of genetic material replication. There is one more problem, very important from the evolutionary point of view - how evolution rate depends on the population size. Unfortunately we were not able to simulate the large

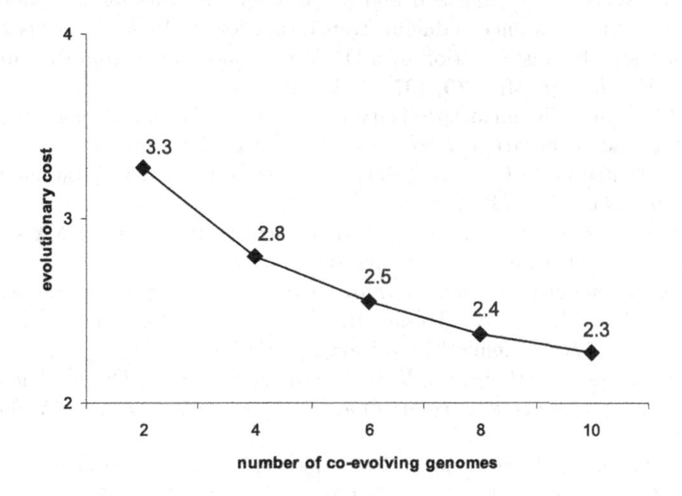

Fig. 1.19. Evolutionary costs for different numbers of co-evolving genomes. Simulations were done for 1 mutation per genome per MCs. Evolutionary costs were estimated by the ratio between the number of eliminating mutations and divergence.

populations of genomes because such simulations need a lot of computer power. Nevertheless, the preliminary studies done with 2 - 10 genomes show that the evolutionary costs decrease with the increase of the population size even if there are no recombination between the evolving genomes (Fig. 1.19). It is not in agreement with some outcomes of the Kimura neutral theory [26]. The neutral theory of Kimura could be now revisited using the extensive computer simulations of the whole genomes evolution.

Acknowledgements

The work was done in the frame of European programs: COST Action MP0801, FP6 NEST - GIACS and UNESCO Chair of Interdisciplinary Studies, University of Wrocław. Calculations have been carried out in Wrocław Centre for Networking and Supercomputing (http://www.wcss.wroc.pl), grant # 102.

References

[1] F. Griffith, Significance of Pneumococcal types, *J. Hyg.* **27**, 113—159, (1928).

[2] O. T. Avery, C. M. Macleod and M. McCarty, Studies on the chemical nature of the substance inducing transformation of Pneumococcus types, I. Induction of transformation by a DNA fraction isolated from Pneumococcus type III, *J. Exp. Med.* **79**, 137—158, (1944).
[3] E. Chargaff, Chemical specificity of nucleic acids and mechanism of their enzymatic degradation, *Experientia* **6**(6), 201—209, (1950).
[4] J. D. Watson and F. H. C. Crick, A structure for deoxyribose nucleic acid, *Nature* **171**, 737—738, (1953).
[5] J. R. Lobry, Asymmetric substitution patterns in the two DNA strands of bacteria, *Mol. Biol. Evol.* **13**, 660—665, (1996).
[6] T. M. Sonneborn, *Degeneracy in the genetic code: extent, nature and genetic implications*, In eds. V. Bryson, H. J. Vogel, *Evolving Genes and Proteins*, pp. 297—377, Academic Press, NewYork (1965).
[7] C. R. Woese, D. H. Dugre, W. C. Saxinger and S. A. Dugre, The molecular basis for the genetic code, *Proc. Natl. Acad. Sci. U.S.A.* **55**, 966—974, (1966).
[8] S. J. Freeland, T. Wu and N. Keulmann, The case for an error minimizing standard genetic code, *Orig Life Evol Biosph.* **33**, 457—477, (2003).
[9] F. H. Crick, The origin of the genetic code, *J. Mol. Biol.* **38**, 367—379, (1968).
[10] Ch. L. Berthelsen, J. A. Glazier and M. H. Skolnick, Global fractal dimension of human DNA sequences treated as pseudorandom walks, *Phys. Rev. A* **45**, 8902—8913, (1992).
[11] S. Cebrat, M. R. Dudek, The effect of DNA phase structure on DNA walks, *Eur. Phys. J. B.*, 271—278, (1998).
[12] P. Mackiewicz, M. Kowalczuk, D. Mackiewicz, A. Nowicka, M. Dudkiewicz, A. Łaszkiewicz, M. R. Dudek and S. Cebrat, Replication associated mutational pressure generating long-range correlation in DNA, *Physica A* **314**, 646—654, (2002).
[13] P. Mackiewicz, A. Gierlik, M. Kowalczuk, M. R. Dudek and S. Cebrat, How does replication-associated mutational pressure influence amino acid composition of proteins? *Genome Research* **9**, 409—416, (1999).
[14] P. Mackiewicz, J. Zakrzewska-Czerwińska, A. Zawilak, M. R. Dudek and S. Cebrat, Where does bacterial replication start? Rules for predicting the oriC region, *Nucleic Acids Res.* **32**, 3781—3791 (2004).
[15] M. Kowalczuk, P. Mackiewicz, D. Mackiewicz, A. Nowicka, M. Dudkiewicz, M. R. Dudek and S. Cebrat, DNA Asymmetry and the Replicational Mutational Pressure, *J. Appl. Genet.* **42**(4) 553—577 (2001).
[16] M. Kowalczuk, P. Mackiewicz, D. Mackiewicz, A. Nowicka, M. Dudkiewicz, M. R. Dudek and S. Cebrat, Multiple base substitution corrections in DNA sequence evolution, *Int. J. Modern Phys. C* **12**(7), 1043—1053, (2001).
[17] M. Kowalczuk, P. Mackiewicz, D. Mackiewicz, A. Nowicka, M. Dudkiewicz, M. R. Dudek and S. Cebrat, High correlation between the turnover of nucleotides under mutational pressure and the DNA composition, *BMC Evolutionary Biology* **1**, 13, (2001).
[18] M. Dudkiewicz, P. Mackiewicz, A. Nowicka, M. Kowalczuk, D. Mackiewicz,

N. Polak, K. Smolarczyk, M. R. Dudek and S. Cebrat, Properties of genetic code under directional, asymmetric mutational pressure, *Lecture Notes in Computer Science* **2657**, 343—350, (2003).
[19] M. Kowalczuk, A. Gierlik, P. Mackiewicz, S. Cebrat and M. R. Dudek, Optimization of Gene Sequences under Constant Mutational Pressure and Selection, *Physica A* **273**, 116—131, (1999).
[20] A. Nowicka, P. Mackiewicz, M. Dudkiewicz, D. Mackiewicz, M. Kowalczuk, S. Cebrat and M. R. Dudek, Correlation between mutation pressure, selection pressure, and occurrence of amino acids, *Lecture Notes in Computer Science* **2658**, 650—657, (2003).
[21] D. Szczepanik, P. Mackiewicz, M. Kowalczuk, A. Gierlik, A. Nowicka, M. R. Dudek and S. Cebrat, Evolution rates of genes on leading and lagging DNA strands, *J. Mol. Evol.* **52**(5), 426—433, (2001).
[22] P. Mackiewicz, D. Szczepanik, A. Gierlik, M. Kowalczuk, A. Nowicka, M. Dudkiewicz, M. R. Dudek and S.Cebrat, The differential killing of genes by inversions in prokaryotic genomes, *J. Mol. Evol.* **53**(6), 615—621, (2001).
[23] M. Dudkiewicz, P. Mackiewicz, M. Kowalczuk, D. Mackiewicz, A. Nowicka, N. Polak, K. Smolarczyk, J. Banaszak, M. R. Dudek and S. Cebrat, Simulation of gene evolution under directional mutational pressure, *Physica A* **336**(1-2), 63—73, (2004).
[24] M. Dudkiewicz, P. Mackiewicz, D. Mackiewicz, M. Kowalczuk, A. Nowicka, N. Polak, K. Smolarczyk, J. Banaszak, M. R. Dudek and S. Cebrat, Higher mutation rate helps to rescue genes from the elimination by selection, *Biosystems* **80**, 192—199, (2005).
[25] D. Mackiewicz, P. Mackiewicz, M. Kowalczuk, M. Dudkiewicz, M. R. Dudek and S. Cebrat, Rearrangements between differently replicating DNA strands in asymmetric bacterial genomes, *Acta Microbiologica Polonica* **52**(3), 245—261, (2003).
[26] M. Kimura, Evolutionary rate at the molecular level, *Nature* **217**, 624—626, (1968).
[27] J. W. Drake, B. Charlesworth, D. Charlesworth and J. F. Crow, Rates of spontaneous mutation, *Genetics.* **148**, 1667—1686, (1998).
[28] M. Ya. Azbel, Phenomenological theory of mortality evolution. *Proc. Natl. Acad. Sci. U.S.A.* **96**, 3303—3307, (1999).
[29] P. M. C. De Oliveira, S. Moss De Oliveira, D. Stauffer, S. Cebrat and A. Pekalski, Does sex induce a phase transition? *Eur. Phys. J. B.* **63**, 245—254, (2008).

Chapter 2

Evolution of the age-structured populations and demography

Agnieszka Łaszkiewicz[1], Przemysław Biecek[2], Katarzyna Bońkowska[2], and Stanisław Cebrat[2]

[1] *Institute of Immunology and Experimental Therapy, Polish Academy of Sciences, ul Weigla 12, Wrocław, Poland,*
[2] *Department of Genomics, Faculty of Biotechnology, University of Wrocław, ul. Przybyszewskiego 63/77, 51-148 Wrocław, Poland, cebrat@smorfland.uni.wroc.pl.*

We describe modeling the population evolution using the Penna model based on Monte-Carlo method. Individuals in the populations are represented by their diploid genomes. Genes expressed after the minimum reproduction age are under a weaker selection pressure and accumulate more mutations than those expressed before the minimum reproduction age. The generated gradient of defective genes determines the ageing of individuals and age-structured populations are very similar to the natural, sexually reproducing populations. The genetic structure of a population depends on the way how the random death affects the population. The improvement of the medical care and healthier life style are responsible for the increasing of the life expectancy of humans during the last century. Introducing a noise into the relations between the genotype, phenotype, and environment, it is possible to simulate some other effects, like the role of immunological systems and a mother care. One of the most interesting results is the evolution of sex chromosomes. Placing the male sex determinants on one chromosome of a pair of sex chromosomes is enough to condemn it for shrinking if the population is panmictic (random-mating is assumed). If males are indispensable for taking care of their offspring and have to be faithful to their females, the male sex chromosome does not shrink.

Contents

2.1. Introduction

Let's start from the end - the end of our life, when our strength weakens and finally the death takes its toll. The mysterious nature of this phenomenon eludes scientific investigations, leaving it still the "unsolved biological problem" [1]. Most of the ageing theories try to describe mechanisms which are suspected to shape the slow decay of life. They invoke various forms of damage to DNA, cells, tissues and organs. The very popular free radical theory of aging may serve as an example [2, 3]. However, evolutionary theories try to reach the roots of the phenomenon, seeking to explain why we age; in other words: how is it possible that senescence, which decreases the fitness of the individual, has evolved at all? Nevertheless, the proper use of these theories requires the awareness of their limitations, all these concepts should be "handled with care", as it was suggested by Le Bourg [4, 5].

One of the first biologists who proposed that ageing is a product of evolutionary forces was Weismann [6]. He believed that there is a genetically programmed death which eliminates older members of population, providing more resources for the younger generation. In this context, ageing represents an adaptation - advantageous for the species, even if it has a negative effect on the individual fitness. Although Weismann changed his views on the evolution of ageing over the course of his life, in the literature his name is associated with the first version of his programmed death theory. In response to this initial idea, Medawar presented his own theory of ageing during an inaugural lecture at University College London in 1951 [1]. He emphasised that the programmed death theory is logically circular; Why do animals age and die? Because they become worn out and decrepit and consequently valueless for the species. But why are they worn out and decrepit? Because they age. In other words, we are not ageless because we

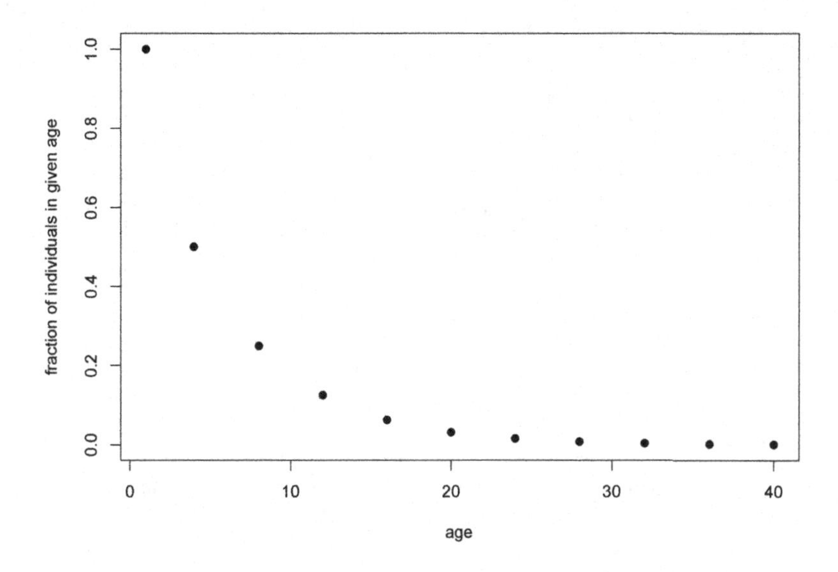

Fig. 2.1. Distribution of number of test tubes of a given age [1]. It is equivalent to the age distribution in populations where individuals do not age and die only because of random accidents.

age. Medawar took another approach; the existence of a post-reproductive period is one of the consequences of senescence; it is not its cause. Developing his model of immortal species, Medawar used as a metaphor random breaking and replacing of test tubes. In a chemical laboratory, equipped with a stock of the test tubes, the broken tubes are replaced by the new ones (age = 0), after some time the number of test tubes of a given age will decline exponentially with age (Fig. 2.1). The older the test tubes are, the fewer there will be of them - not because they become more vulnerable with increasing age, but simply because the older test-tubes have been exposed the more times to the hazard of being broken [1].

Avoiding Weismann's trap, Medawar assumed the equal probability of reproduction for young and old individuals. Since a non-ageing individual tends to have statistically the same number of progeny, every year, from puberty to death, the total number of produced progeny increased linearly with age.

These two functions: exponential decrease of the number of individuals with age and linear increase of progeny with age shape the contribution of

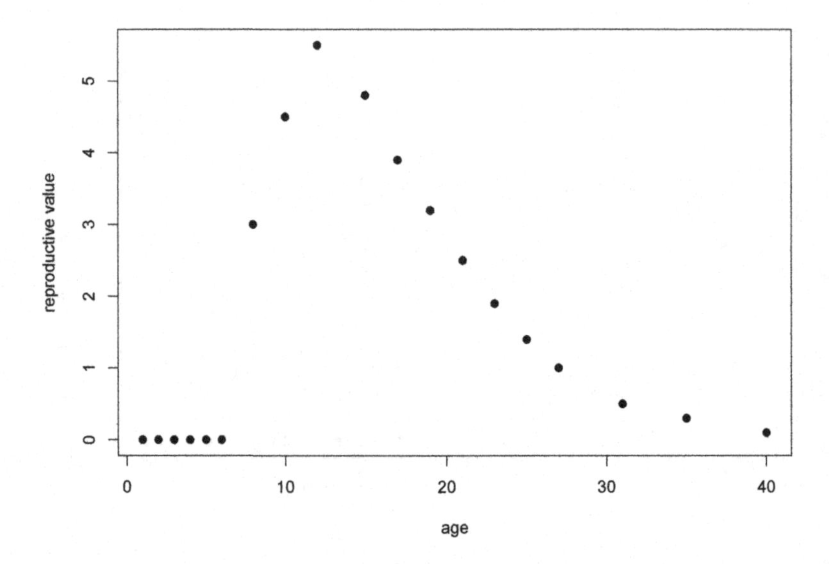

Fig. 2.2. Reproduction values of different age groups according to Goldsmith [10]. The youngest organisms, before the minimum reproduction age do not contribute to the reproduction. The contribution of the older groups is lower not because they are less fertile but because there are less of them in population.

each age group to the total reproduction potential of population. As it can be seen in Fig. 2.2, the reproductive value of the group of older individuals is low. In Medawar's words: "The older the age group, the smaller is its overall reproductive value... This is not because (the test tubes) of the senior group are individually less fertile but merely because there are fewer of them; and there are fewer of them not because they have become more fragile - their vulnerability is likewise unaltered - but simply because, being older, they have been exposed more often to the hazard of being broken" [1].

Summing up, Medawar's idea states that the population loses individuals with time because of the random death, so the reproductive effect of the group of older individuals decreases with age, therefore the selection for their continued survival and reproduction should also weaken. Since these verbal arguments have been formalized, the idea began life as a mathematically coherent theory [7, 8]. It turns out that before the onset of reproduction, the force of natural selection is highest and constant. After this time, however, the force of natural selection progressively falls, reaching zero around the cessation of reproduction. As a consequence, the genes beneficial early in life are favored by natural selection while genes which should be expressed and necessary for surviving the late ages, after the reproduction period, can freely accumulate deleterious mutations.

According to mutation accumulation theory proposed by Medawar himself, mutational pressure introduces new hereditary factors, which could lower the fertility or viability of the organisms. However, if they are expressed late enough, the force of selection will be too attenuated to oppose their establishment and spread [1]. In this so called "selection shadow" many germ-line mutations neutral for young individuals, but with late deleterious effects accumulate passively over generations leading to senescence and death [9]. According to presented theory, ageing is a nonadaptive trait, being only a by-product of the way natural selection operates. It claims that senescence evolved because it is out of reach of natural selection and there was no possibility to eliminate it. That is why these theories are sometimes called the law of unintended consequences. However, the scientific opposition believe in "Death by Design" and try to revive Weisman's idea of ageing evolving as an adaptation. They criticise Medawar's model of non-ageing species, which in their opinion is oversimplified and ignores the evolutionary importance of older animals. They claim that the probability of reproduction of old non-ageing individuals is not lower than that of young ones (Weisman's trap), nor is it equal (Medawar's idea); in fact the old immortal individual has a higher probability of reproduction, since it is more mature and better fit. Moreover, older non-ageing animals have lower probability of death than younger less experienced individuals: "The old king is less likely to die in the war than the young foot soldier" [10]. Using this kind of arguments "opposition" is trying to discredit Medawar's assumptions and call reliability of his model into question.

Finally, what is certain about the end of our biological existence? The end is certain.

2.2. The Penna model

2.2.1. *Description of the standard Penna model*

In this chapter we will present the results of computer simulations of populations' evolution based on the Penna model [11]. Hundreds of papers using this model or citing it have been already published. Authors of these papers (including Penna himself) claim that it is an ageing model and very often it is suggested that the model assumes the Medawar hypothesis of ageing [12]. In fact, the Penna model does not assume the Medawar hypothesis but the results of simulations using this model support the Medawar hypothesis. Moreover, the Penna model could be very useful in describing many phenomena connected with the evolution of biological populations, like altruism [13], menopause [14], the role of the immunological system [15], of mother care [16], even the evolution of sex chromosomes [17]. There are two main versions of the model, haploid and diploid. We will describe the diploid version only, because it fits much better to the biological reality. In the description of the model we will try to introduce some genetic terminology just to show the correspondence between some genetic elements and mechanisms and the virtual world of the model. It would facilitate further contacts of non-biologist readers with biologists.

In the model, populations are composed of individuals, each one represented by its genome composed of two haplotypes which are the bitstrings L bits long. Bits represent genes. Bits placed in the corresponding position (locus/loci) in the bitstrings are called alleles. If a bit is set to 0, it represents a wild form of an allele (a correct one), if a bit is set to 1, it represents the defective allele. If one allele is defective and the other one in the corresponding locus of the second bitstring is wild, then two situations could be considered: if the defect is dominant - the function (or phenotype) is defective, if the defective allele is recessive - the function of the locus is normal, if both bits at the same positions are defective, the phenotype is always defective. In all presented here simulations we have assumed that all defective alleles are recessive. The main assumption of the Penna model is the chronological switching on the pairs of alleles; after birth, in the first Monte-Carlo step (MCs), the alleles at the first locus are switched on, in the second step, the alleles at the second locus and so on. One MCs corresponds to a time unit/period. The number of defective loci which kill the individual is one of the model parameter called the threshold T. Death of individual caused by the expression of the T phenotypic defects is called the genetic death. If the individual expresses all pairs of alleles in its two bitstrings, then even if it does not reach the threshold T it is going to die because of reaching the "absolute maximum age" L. In fact, when parameters are properly set, individuals do not reach that age, they are dying earlier because of the genetic death.

When an individual reaches the minimum reproduction age R (after switching on R pairs of alleles) it can reproduce. In all our simulations we assume the conditions of sexual dimorphism and we declare that the sex of a newborn can be set as a male or a female with an equal probability. A female at the reproduction age produces a gamete. To do that the diploid genome of that female is copied and during this process a new mutation is introduced into each haplotype (a bitstring) at the randomly chosen position with a probability M/L per loci (thus M is the average number of mutations per haplotype). If the bit chosen for mutation has a value 0 - it is changed for 1, if it is already 1 it stays 1 - there are no reversions. The two bitstrings recombine in the process mimicking the crossover with a probability C. The position, where the two bitstrings are cut for the crossover is randomly chosen. One of the two products of the crossover is randomly chosen as a female gamete and then, the male individual at the reproduction age is chosen randomly from the whole pool of "adult" males in the population. This male produces its gamete in the same way as a female has done it. The female and male gametes form the diploid genome of an offspring, its sex - male/female is set with an equal probability. In the next MCs its first locus will be checked for the genetic status of its bits (genes). In one MCs an age of all individuals increases by one and each female at the reproduction age can give a birth to a new baby with a probability B or can give a birth of B new babies. There is another mechanism controlling the birth rate in the model. It is a logistic equation of Verhulst: $V = 1 - N_t/N_{max}$, where N_t is the number of individuals in the population in the t-step of simulation, N_{max} is a maximal size of the population called also a capacity of the environment and V is a probability that a new offspring will survive its birth.

In fact there are only seven parameters in the standard Penna model:

- L - the number of loci,
- M - the average number of mutations introduced into the haploid genome during the gamete production (usually M=1 per haploid genome per generation),
- B - the number of offspring produced by each female at reproduction age at each time step or the probability of giving the offspring,
- R - minimum reproduction age,
- T - the upper limit of expressed defects, at which an individual dies,
- C - the probability of a crossover between parental haplotypes during the gamete production,
- V - the Verhulst factor $V = 1 - N_t/N_{max}$.

2.2.2. *Results of standard simulations*

There are many possibilities of starting the simulations. Usually we start with randomly generated population of half of N_{max} size with perfect genomes (all bits set for 0). Randomly generated means here evenly distributed age of individuals and equal numbers of females and males. During the simulation, a mutational pressure introduces mutations into haplotypes while a selection tends to eliminate them. Figure 2.3 shows the distribution of defective genes in the genomes in equilibrium. Among the genes expressed before the minimum reproduction age (R) the fraction of defective alleles is the lowest. After R the fraction of defective genes grows with the age of their expression, eventually reaching 1 for the late ages which means that there are no functional genes in the whole genetic pool which could be necessary for surviving during those late periods of life. Such a distribution of defective genes determines the age structure of populations. The mortality of the youngest individuals is the lowest and it is growing with age. The position of the first genes which are set in the whole population for 1 determines the maximum life span of individuals in the population. More precisely, the maximum life expectancy is (T-1) steps longer than indicated by the position of the first locus set to 1.

Note that in these populations individuals can reproduce with equal probability to the end of life. Thus, the results support the Medawar hypothesis of ageing assuming that even if individuals can reproduce to the end of their life, their impact on the total reproduction potential of population decreases and selection does not care about their genetic status allowing the accumulation of mutations in genes expressed during the late periods of life.

2.2.3. *The role of parameters in modeling the age-structured populations*

2.2.3.1. *Random death*

One of the most important factors in the population simulations is the method of keeping the stable size of populations. There are two factors in the Penna model: birth rate B and Verhulst factor V responsible for that. The birth rate has to be set high enough to secure at least the replacement of fraction of dying individuals in the evolving populations. A higher birth rate leads to the roughly exponential growth of population and that is why the logistic equation of Verhulst is introduced. Deaths due to

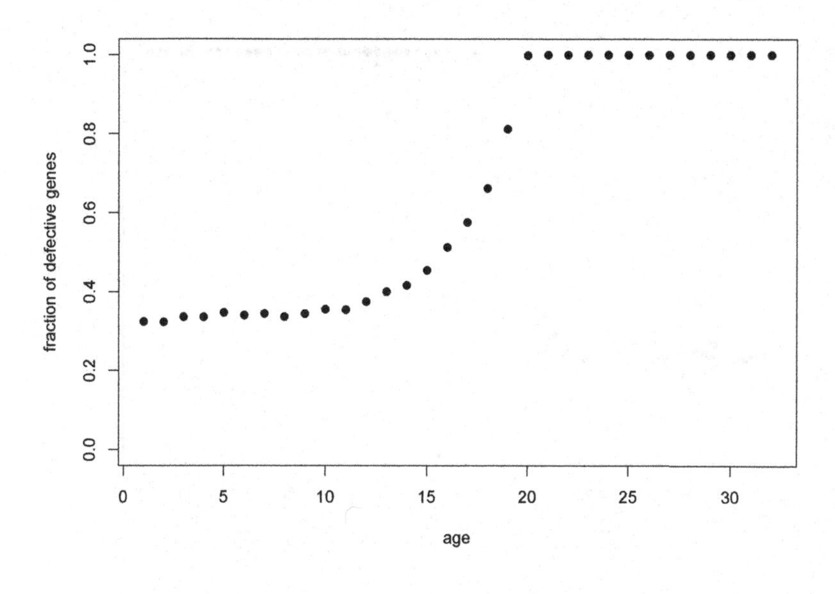

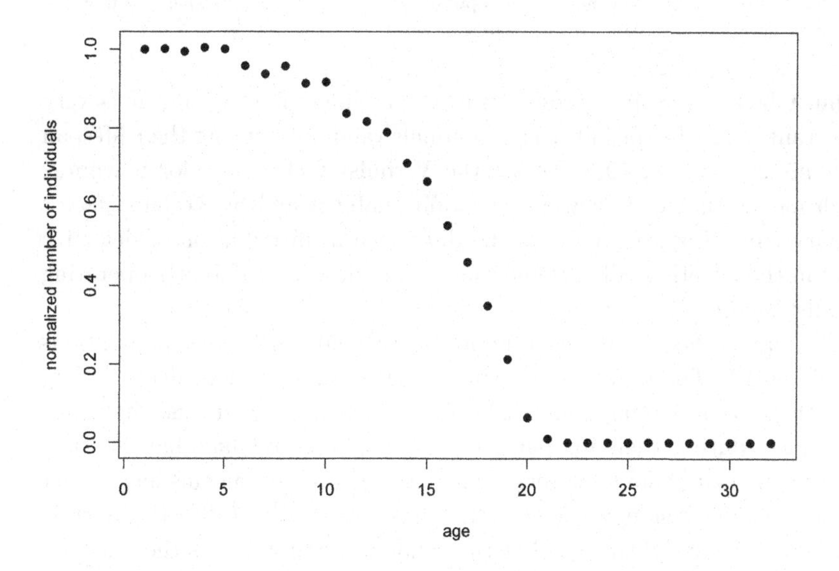

Fig. 2.3. Distribution of defective genes (bits set to 1) in the genetic pool of virtual Penna populations - top panel, and age distribution of the corresponding population - lower panel. The age corresponds to the number of MC steps and to the number of bits switched on. Parameters of simulations: L=32, M=1, C=1, B=1, T=3, R=8.

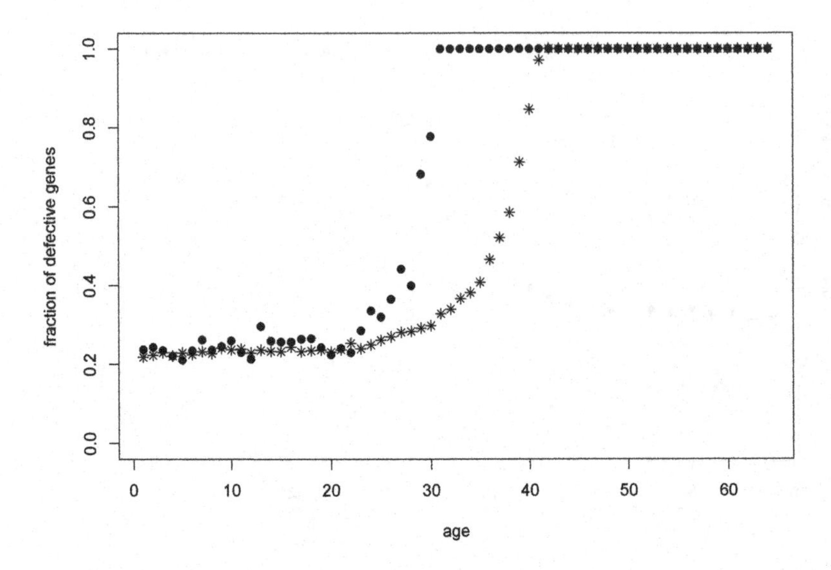

Fig. 2.4. The difference in distribution of defective genes in the genetic pool of populations evolving under Verhulst factor operating at birth only (stars) or "killing" all individual independently of their age (circles).

Verhulst factor may be introduced in a few different ways and it is very important if this factor kills the individuals randomly during their lifespan or at birth only [18, 19]. We set the Verhulst factor only for newborns which means that each newborn is additionally tested by Verhulst factor for surviving. Figure 2.4 shows the difference in distribution of defective genes in the genetic pool of populations evolving with differently operating Verhulst factors.

One can wonder why the random death could affect the populations so differently. There are two reasons. One is that random death killing at birth kills individuals independently of their genetic status. If it kills individuals during their lifespan then, the older individuals have to pass the tests of their genomes several times. These older organisms have larger fraction of their genomes correct which has been testified directly by their age. The other conclusion is that the random death weakens the selection for genetic status of individuals. In fact, in Nature such a random death operates mainly at very early stages of development or at very late periods of life.

2.2.3.2. *Threshold T*

The average life expectancy, which supposes to mirror the population's health, rose dramatically during the 20th century. The most striking example is East Asia, where the life expectancy at birth increased from less than 45 years in 1950 to more than 72 years nowadays (it means that it was increasing by six months each year!). So far, however, Japanese women enjoy the longest expectation of life: it approached 85 years in 2003, in comparison to 63 years in 1950 and 44 years in 1900 [20]. This outstanding achievement of our civilization becomes a challenge. Combined with the decrease in fecundity (number of children born by one woman), it has brought the global ageing of the human population and fears of pension crises. Such deep demographic changes strongly influence the economic, educational and social policies. Therefore, there is an urgent need for correct forecasts of expectation of life for future decades. Unfortunately, there is no agreement among scientists in these predictions.

How much can the human life span be extended?

This question could have many contexts. It can be understood more philosophically as the quest for the truth about the possibility of physical immortality. In other words: whether we are biologically designed to die and nothing can be done about it or rather, putting stop to ageing is only the matter of time and eventually human beings will live forever, unless killed by random death. More practical, however, this is a question about the demographic trends in the coming decades, and there are at least two schools which aspire to answer it. They are characterized by different points of view on current human populations' demographic trends and predict different future demographic scenarios.

Adding life to years...; The first idea, developed by Fries [21], is rooted in the conviction that there is an inherent limit imposed on the human longevity by biology, which determines the maximal life span. Consequently, there must also be a ceiling to human life expectancy. Fries even assessed the position and shape of the "ideally rectangular survival curve", which would be the result of the hypothetical elimination of all premature deaths, and set ultimate life expectancy level at 85 years. The intuitive consequence of this process of rectangularization should be the so called compression of morbidity. If the onset of infirmity is postponed (*e.g.* by changes in lifestyle or improvements in medicine), the period with the debilitating effects of illness or disability reduces, because of the fixed maximum age. This concept predicts that increasing life expectancy should not result in the extension of morbidity: in "ever older, ever more feeble, and ever more expensive-to-care-for populace" [21], but rather in the compression of infirmity into a short period prior to death, associated with the reduction in the scale of health care systems.

Summing up, the presented interpretation of demographic changes may be illustrated using a popular catch-phrase as a process of "adding life to years". There is no place for prophets of physical immortality in this scenario.

...or years to life? More and more scientists come to believe that this theory does not properly describe the demographic reality. Oeppen and Vaupel [22] presented a comparison between the forecasts done in the past and real-life development as a story about breaking all the limits predicted by demographers to life expectancy: from 64.75 years calculated as its ultimate level by Louis Dublin in 1928 to 85 years set again by Olshansky *et al.* in 1990 [23]. The main message of their article is that there are no signs that life expectancy is approaching any limit and that the belief in such limits led to underestimation of life expectancy in the past and definitely will not help in forecasting the future trends. This observation does not discredit the rectangularization hypothesis itself. As long as the increase in maximum age at death is slower than the increase in average age at death (life expectancy) the survival curve will show rectangularization. But such rectangularization without the belief in the fixed maximum age will not provide us with such a simple and precise answer to what we can expect in the close future as it was in Fries' scenario.

And reality seems to be even more complicated and difficult to forecast. As a matter of fact, the pattern of changes in the survival curve that occurred in the first half of 20th century did show rectangularization. But scientists report that this trend has been replaced in some countries by a near parallel shift of the curve to the right [24]. This leaves the future wide open for speculations.

More generally, some scientists stress the significance of medical interventions and public health as the determinants of mortality changes; others see their origin mostly in economical, social, cultural and behavioral factors. Nevertheless, one century is a too short period for any significant reconstruction of the genetic pool of the human population, and thus in broad terms the observed mortality transition must have been driven by the changes in the individual/environment relation and thus by the increase in "threshold value T", whatever the detailed scenario of this process was. Consequently, it should be possible to simulate the changes in the human population just by increasing the individuals' tolerance to the number of expressed defective characters.

Since the T parameter in the Penna model describes the individuals' tolerance to the deleterious traits that they suffer from, we used it to model the dynamic of demographic changes [25]. When T is set for one, it means that every single deleterious phenotypic trait kills the individual. The higher the T value means that each individual copes better with its genetic load. In the real world, this value might correspond to the standard of medical care, lifestyle, diet, hygiene, education and all other factors that help an individual to escape some fatal effects of his genetic traits. Therefore, this is a very capacious parameter and it can be generally interpreted as the factor in charge of the relation between individuals and environment. Notice that changes in the age structure of human populations were observed in very short period - in Japan it was only 50 years - less than two generations. There was no time for rebuilding the genetic pool of populations. Thus, all changes suppose to correspond to relations between the environment and populations. That is why we used for these studies the populations in the equilibrium and then we have changed the values of T parameter [25].

The plots in Fig. 2.5 show the role of changes of the T parameter values. In these series of simulations populations evolved under the values of parameters as follows: L = 640, R = 200, B = 0.25, N_{max} = 50 000, C = 1. Populations were allowed to evolve for 80 000 MC steps under the threshold $T = 3$. Then, one million zygotes were produced and the life span of each zygote was anticipated without further evolution, just on the basis of their genetic load. This was repeated for different values of parameter T. When these values were not integers, the probability of surviving under a given T was equal to a fractional part of T, which means that individuals under threshold $T = 3.3$ had a probability 0.7 to die when three bad mutations were reached. In this way the anticipated age distributions of individuals for different thresholds T were prepared and plotted in Fig. 2.5. The increase of the threshold T value leads to the rectangularization of the curve describing the age distribution of the population with relatively small shift of the maximum age to the right. Comparison of these results with the data from the real Swedish population shows that the increase in the T value can mimic the demographic changes in the age distribution of the human population. Data presented in Fig. 2.6 are based on those presented in Fig. 2.5 but the x-axis has been rescaled to fit the human life span; the period from 20 - 100 years in the human life span corresponds to the period from minimum reproduction age $R = 200$ to the maximum age in simula-

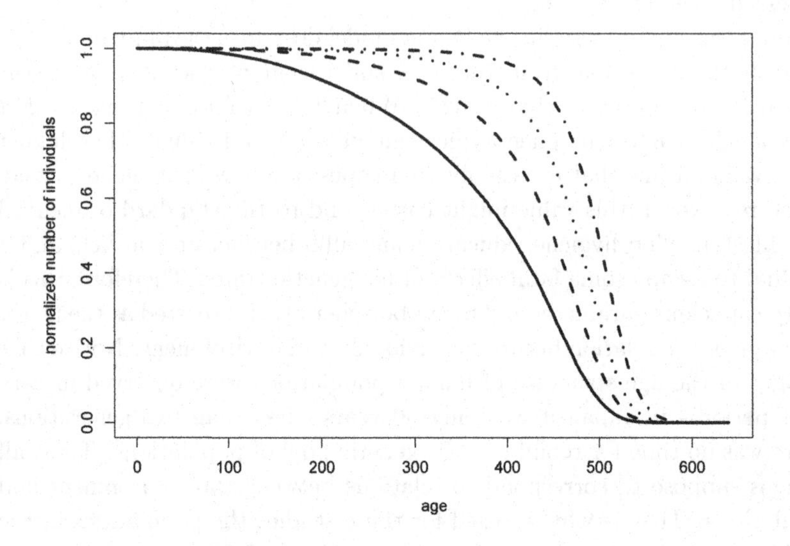

Fig. 2.5. Effect of changes of the T parameter values on the age distribution of populations. Parameters of simulations of the original population: L=640, R=200, M=1, C=1, B=0.25, T=3. After 80 000 MCs under threshold T=3 the anticipated age structure of such population under higher T value was calculated; T=3, T=4, T=5, T=6.3 from lower left to upper right curves. For details see the text.

tions. Plots generated by simulations fit very well the age distribution of human populations of second half of the 19th century and part of the 20th century [25]. However, for the end of the 20th century and for the latest data rectangularization shown by simulations seems to be too strong. The model renders very well the level of mortality in the middle ages but the mortality of the oldest individuals is overestimated. Interestingly, demographers claim that life expectancy at older ages relatively stable so far has increased at a spectacular pace just since the second half of the 20th century. This observation leads to the hot debate about the tail of the human mortality curve.

Probably the mortality curve would be better represented if the gradient of the number of expressed genes per year was introduced. This means that a higher number of genes would be expressed in younger ages and progressively this number would decrease in older ages. Such a modification seems to be biologically legitimated and it would be very interesting to test it by simulations in the future. An interesting continuation of this work is

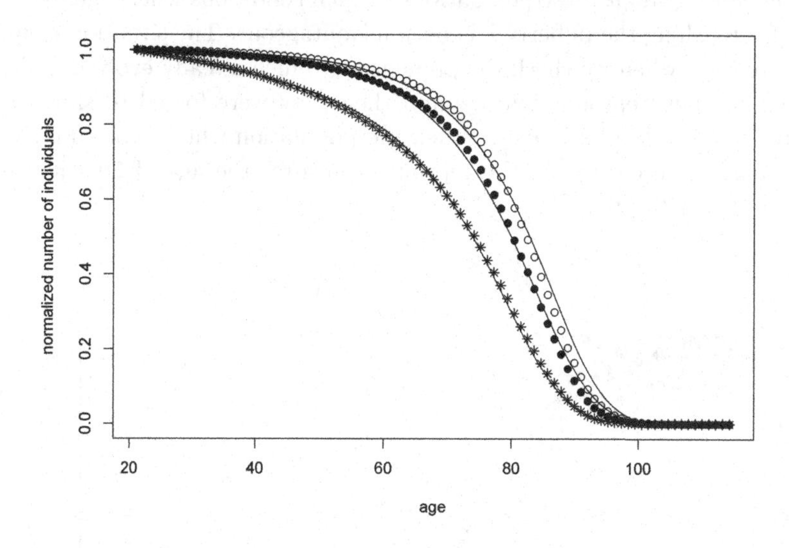

Fig. 2.6. The age distribution of the Swedish populations in different periods (1930-39, 1980-89 and 2000-2003 - lines from lower left to upper right) and corresponding results of simulations under different T values: 3.8, 4,9 and 5.4 symbols from lower left to upper right. For more details about rescaling the x-axis values see the text.

the modeling of the increase in the T value coupled with the decrease in parameter B. Such an approach was used by to study the effects of the dramatic decrease in the birth rate observed in the developed countries on the age structure of the human population [26].

Summing up, increase in parameter T value leads to the rectangularization of the survival curve with relatively small increase in the maximal age, reflecting very well these general demographic trends observed in the human populations. However, how it influences future mortality trends remains a matter of speculations.

Shift of the minimum reproduction age

Some peoples suggest that the best way to increase the human life span is to shift the minimum reproduction age for the later periods of life. Of course it is a good idea, which is shown in Fig. 2.7. When the simulations are started with the increased reproduction age - R the population shifts the whole reproduction period to the older ages and the increase in the life expectancy corresponds roughly to the shift of R and shift in the distribution of defective genes (Fig. 2.8). Nevertheless, one has to remember that in this series of simulations populations evolved under higher R, thus evolution adapt these populations to such conditions where the longer time for reaching the puberty has been advantageous. The situation is not as optimistic, when we check the populations which already evolved under lower minimum reproduction age and then they were forced to shift the reproduction to the later ages. When the population which evolved under R = 5 was forced to start the reproduction after the age of 50 - it was extinct (Fig. 2.9) [27].

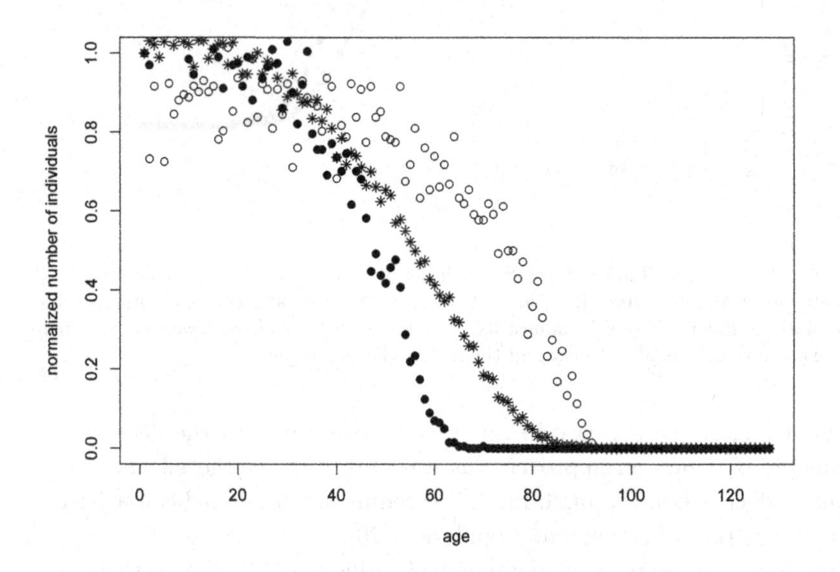

Fig. 2.7. The effect of the increased minimum reproduction age on the age distribution of population. Populations evolved under different R parameter values: 5 - filled circles, 20 - stars, 50 - open circles. The rest of parameters: L = 128, M = 1, C = 1, B = 1, T = 3.

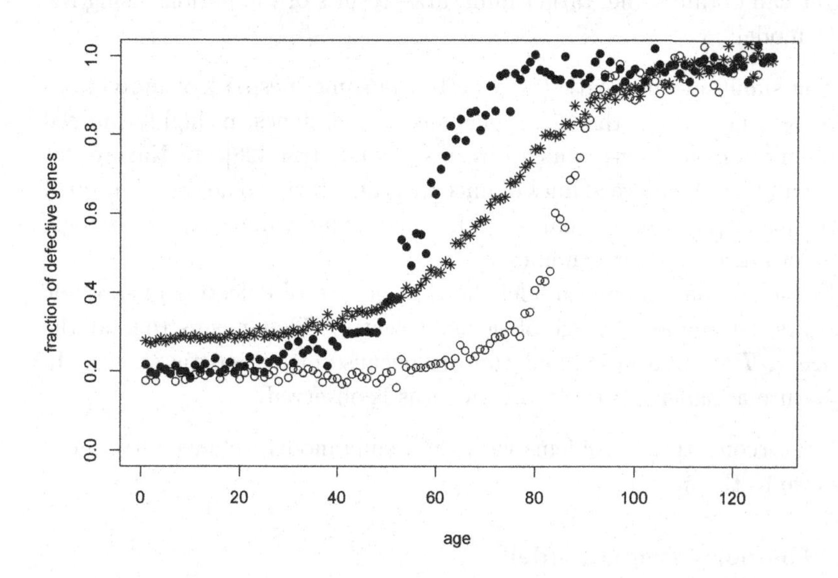

Fig. 2.8. The effect of the increased minimum reproduction age on the distribution of defects. Populations evolved under different R parameter values: 5 - filled circles, 20 - stars, 50 - open circles. The rest of parameters: $L = 128$, $M = 1$, $C = 1$, $B = 1$, $T = 3$.

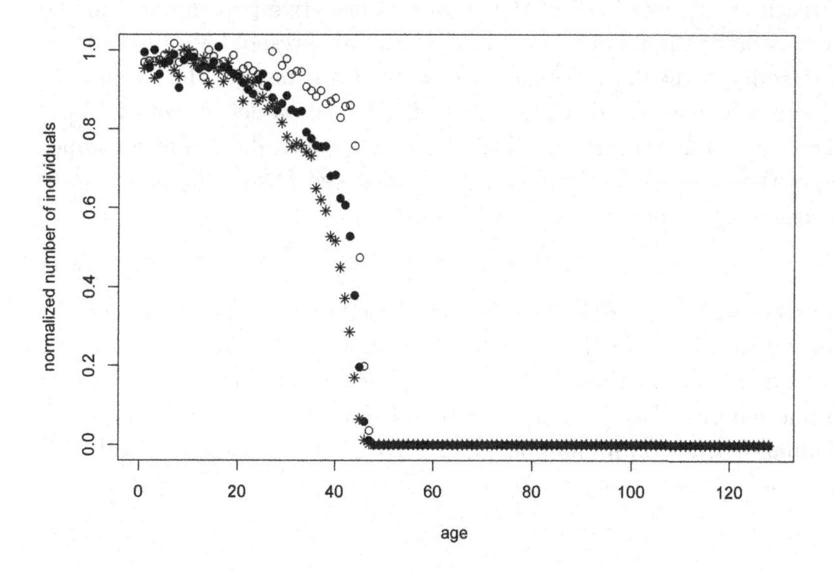

Fig. 2.9. The effect of evolution of population forced to change the age of reproduction to 10 - stars, 20 - full circles, 40 - open circles, 50 - population extinct.

Penna with noise

One can obtain some, rather unnatural, results of simulations using the Penna model;

- The simultaneous death (or exactly the same lifespan) of individuals possessing exactly the same genomes *i.e.* in clones, in highly inbreed populations or twins, which obviously is not true [28]. In Nature, for example in the inbreed lines of mice we still observe some distribution of the life spans of individuals - the life span of individuals is not precisely determined by their genomes.
- In the standard Penna model the threshold T of defective phenotypic traits determines the age of "genetic death". That means that at the age $< T$ no one individual can die because of its genetic status. In Nature a higher mortality of newborns is observed.

To overcome these problems with the Penna model we have introduced the noise to the model.

2.3. The noisy Penna model

For the "noisy extension" of the model, the diploid sexual Penna version has been used. In our version of the standard model, the Verhulst factor controls the birth-rate and there are no random deaths of organisms later during the lifespan. Individuals die only because of the genetic death when they reach the threshold T of the expressed defective phenotypes. In the noisy version of the model, there is no declared threshold T. Instead, we have introduced the fluctuations of the state of organisms. The variance of fluctuations increases with the number of switched on defective loci [15].

The state of an individual i is denoted $I_i(t) \in \mathcal{R}$ is defined as a composition of the inner state of the individual (denoted $P_i(t) \in \mathcal{R}$) and a state of environment (denoted $E(t) \in \mathcal{R}$). Thus

$$I_i(t) = E(t) + P_i(t), \tag{2.1}$$

where $E(t) \sim \mathcal{N}(\mu_{E(t)}, \sigma_E^2)$ corresponds to the fluctuations of environment in time t while $P_i(t) \sim \mathcal{N}(\mu_{P_i(t)}, \sigma_i^2(t))$ corresponds to the inner fluctuations of the individual i in time t. In the simplest case, the expected value of both fluctuations is $\mu_{P_i(t)} = \mu_{E(t)} = 0$ and the variance of the state of an individual depends on its number of defective loci $g_i(t)$ expressed till time t, *i.e.*

$$\sigma_i^2(t) = \sigma_0^2 + g_i(t)\sigma_d^2. \tag{2.2}$$

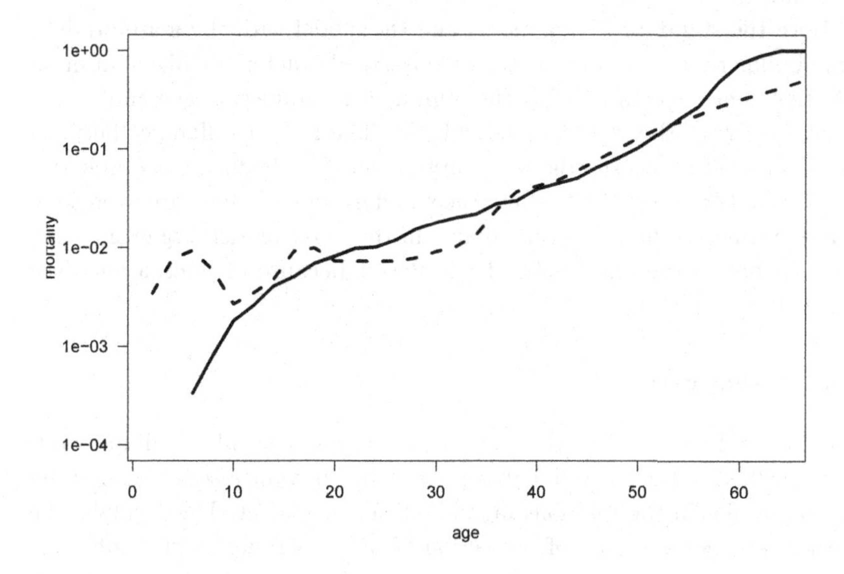

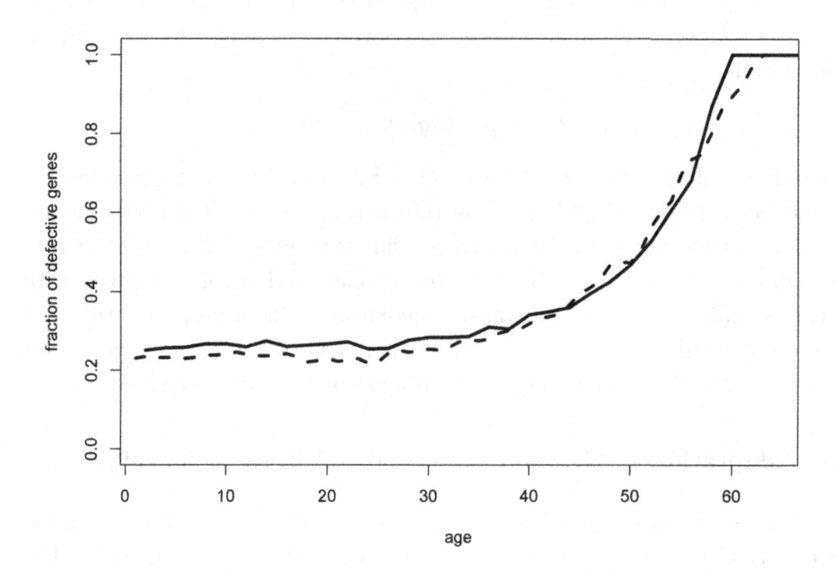

Fig. 2.10. The mortality curve (upper plot) and frequency of defective genes (lower plot) in the standard Penna model (solid line) and in the noisy Penna model (dashed line). Note the logarithmic y-axis of the mortality curve, x-axis is scaled in the arbitrary age units.

One may modify $\mu_{E(t)}$ and include a drift or seasonal changes in the environment.

Both the standard Penna model and the model with the noise produce very similar results. In both cases we observe characteristic distribution of defective genes expressed after the minimum reproduction age and a very low mortality of the youngest individuals. The largest difference between the two models concerns the mortality of individuals during the first two time units. For threshold $T = 3$ in the standard model, there are no genetic deaths during the first two time units. In the noisy model, organisms may die even before the expression of any defect because of fluctuations (see Fig. 2.10).

2.4. Mother care

In the noisy Penna model, the state of an environment affects all individuals regardless of their age but it is possible to introduce some biologically legitimate bias in the relations individual - environment. For example, the mother care as some kind of protection of the babies against the influence of fluctuations of the environment during the first periods of their lives affects these relations significantly. To model the mother care, the effect of the states of the environment on the newborns is decreased by (see [16] for more details)

$$\rho(i,t) = 1 - exp(-(age(i,t) + 1)/\lambda_{MC}).$$

The individual dies if $\rho(i,t)E(t) + P_i(t) > F$. After λ_{MC} steps the effect of the mother care is negligible while in the early stages of life it is significant.

We call this "mother care" to stress that this effect influences the very first periods of life but it could be the proper feeding of newborns with mother's milk as well as an intensive neonatal medical care. In Fig. 2.11 we present results for $\lambda_{MC} = 4$. The distribution of defects is similar for both models and the mortality curve differs only in early stages of life.

2.5. Adaptation to the environmental conditions - learning

In the noisy Penna model, the variance of fluctuations of an individual state is a sum of its inner noise and the noise of an environment. The impact of these two components on the state and evolution of population is different. The personal component is independent for each individual while the environmental noise is the same for each individual. That is the reason why the reactions of individuals for the environmental fluctuations are diversified. In the version of the noisy Penna model presented above, the fluctuations have Gaussian distribution with the average $\mu_{E(t)}$. Now, we introduce a signal into the expected state of the environment. The signal $\mu_{E(t)}$ is a periodical function with the period D, thus $\mu_{E(t)} = \mu_{E(t+D)}$. Individuals know that the signal is periodical, and are equipped with a mechanism of learning the signal. They estimate components of the signal by weighted average of the state of the environment in survived periods. In the more formal way, the learning mechanism affects the expected state of individual fluctuations

$$\mu_{P_i}(t) = \sum_{j=1}^{\infty} L(i, t - j * D) w_j E(t - j * D)$$

where $L(i,t) = 1$ if individual i lived at time t and 0 otherwise while weights w_j are

$$w_j = e^{-(j-1)/\lambda} - e^{-j/\lambda}.$$

This adaptation mechanism allows to reduce the mortality in case when an individual have learned the periodical signal. Results for different λ are presented in Fig. 2.12. It is also observed in real populations that mortality of newborns is higher than of a bit older individuals. The results depend on the maximal signal value $\mu_{E(t)}$ and do not depend on the form of the periodic function, thus results for constant $\mu_{E(t)} = A$ are similar to those obtained with $\mu_{E(t)} = A\sin(t/\pi)$ (results not shown).

The next question is how the intensive protection of newborns against environment fluctuations could influence their mortality during the later periods of life. It is rather well known effect that children who are very strongly protected against any infections during the first periods of life and live in almost sterile conditions are more vulnerable for infections later.

Plots shown in Figs. 2.13-2.14 indicate, that one could really expect slightly higher mortality of young individuals if they are isolated from environment influences during the very early periods of life.

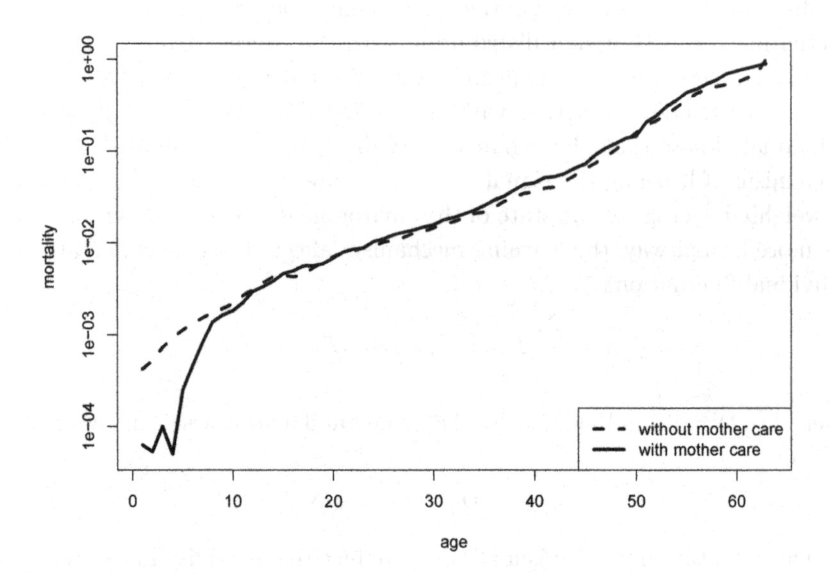

Fig. 2.11. Mortality curves for the population with and without mother care. There is no learning mechanism, thus in early years the mother care protects newborns and after this period the mortality with and without mother care is similar.

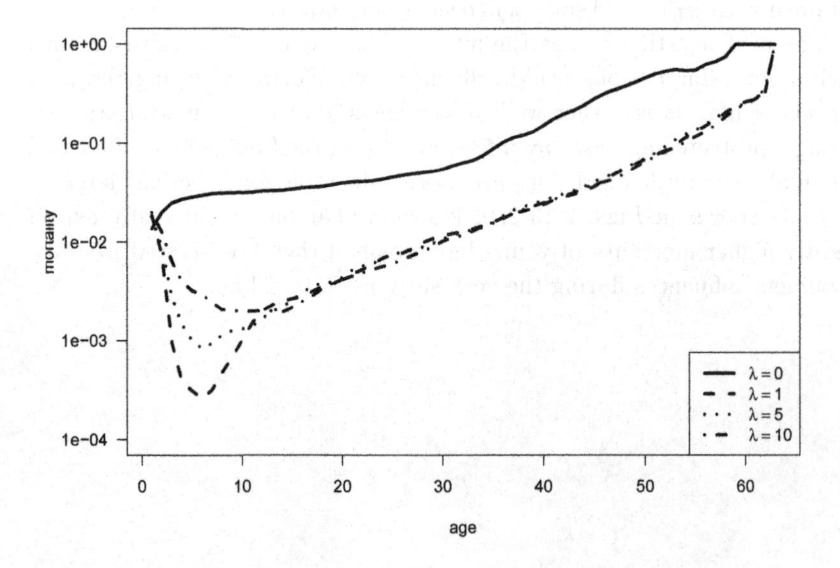

Fig. 2.12. The mortality curves for different learning coefficients λ.

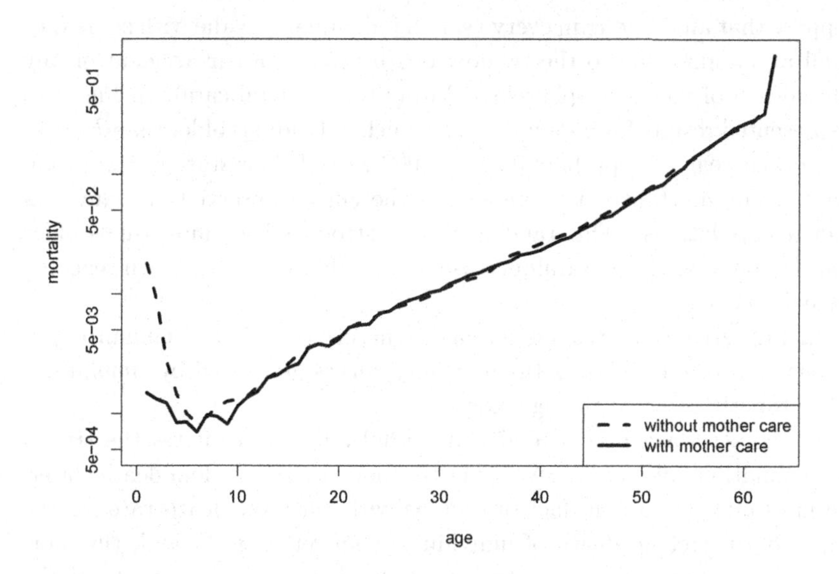

Fig. 2.13. The mortality curve for learning coefficient $\lambda = 5$ with and without mother care.

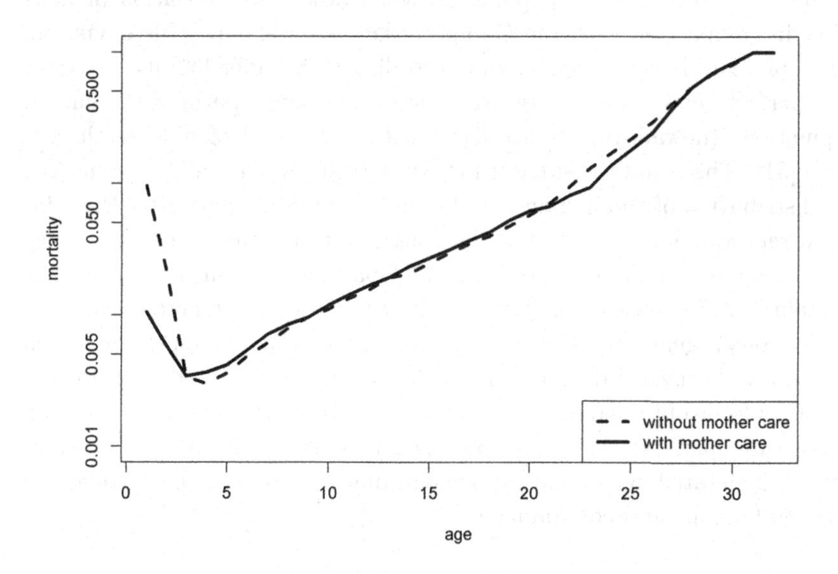

Fig. 2.14. The mortality curve for populations with and without mother care. For early periods of live the mortality of newborns is lower in case of mother care, but then it is a bit higher.

2.6. Additional risk factors

Suppose that after age H in every year an individual may die with some very small probability due to the random death (*e.g.* in a car accident or any other death of young people released from the parental care). In Fig. 2.15 we presented results for populations, in which individuals older than $H = 12$ die in each year with probability $p = 0.002$ even if their state is lower than the F. The mortality curve resembles the curves observed in many real human populations. This random death introduced for individuals older than 12 could be also natural increase in mortality connected with reaching the puberty age.

In Fig. 2.16 we presented the age structure of different human populations and for comparison, the mortality curves generated by simulations with properly rescaled the age axis.

In the standard Penna model, if Verhulst factor regulates the size of the population only at the level of birth, there are no random deaths later, the mortality curve is in fact of s-shape with very low death rate for the youngest (in fact no death of individuals younger than T) and, the mortality rate above the Gompertz law prediction for the last parts of the lifespan. In our model the shape of the mortality curves changes. Even the youngest organism may die, and the results of simulations better fit to the real mortality curve of human populations.

For the oldest part of populations some downward deviation of mortality in comparison with the Gompertz law is observed. This deviation, called plateau, is controversial and according to Stauffer [30] it is a result of imperfect demographic data available for the oldest parts of the human populations (nowadays it concerns data from the end of nineteenth century) [31]. The results of standard Penna model simulations reproduce the age distribution of the human population in its middle part after the minimum reproduction age, but not the oldest part of populations. The noisy model enables simulation and analysis of parameters which influence the mortality of the youngest individuals. Nevertheless, if the heterogeneity of populations is generated by introducing specific parameters into the Penna model, the downward deviation from the Gompertz law prediction for the oldest could be obtained in the computer simulated populations [15, 32, 33]. There is an open question if the heterogeneity observed in the human population is generated by noise in the demographic data or noise in the relations between humans and environment.

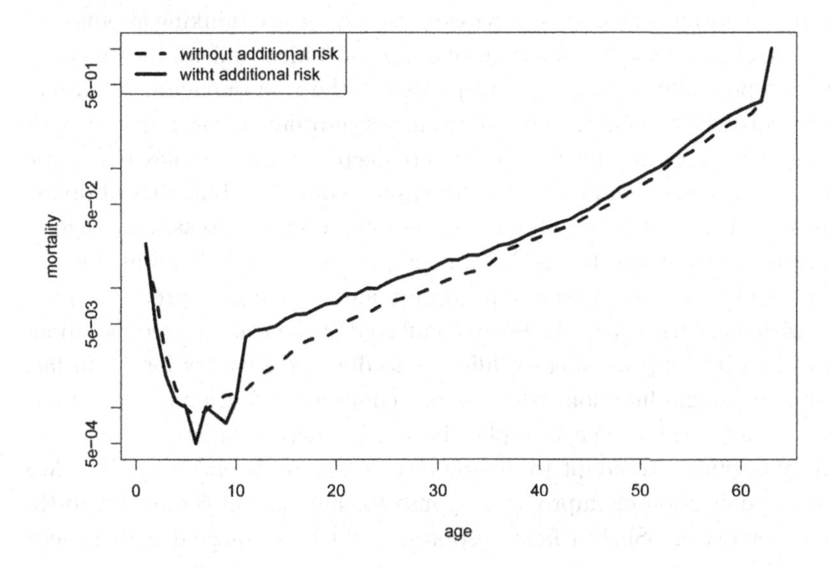

Fig. 2.15. The mortality curve for the population with additional risk of death after $H = 12$ years.

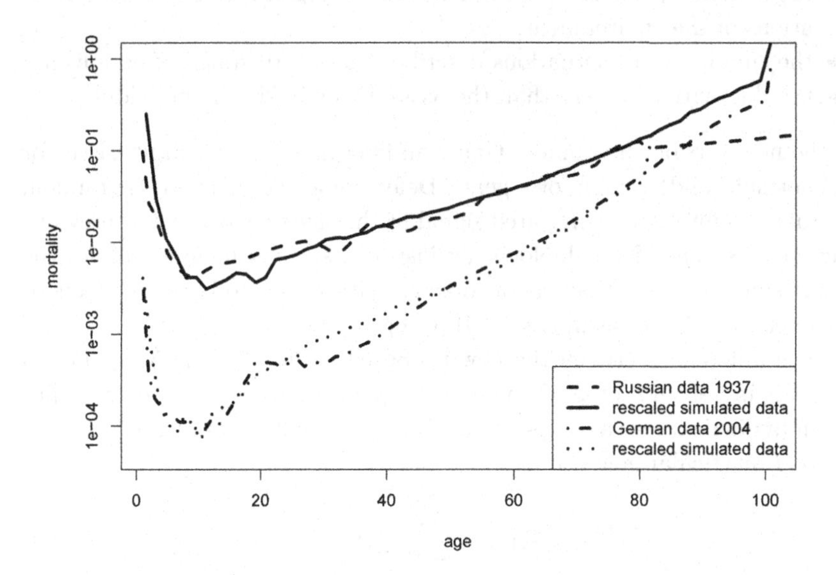

Fig. 2.16. The age structures (upper plots) and mortality curves of different human populations and populations generated in simulations with the extended noisy Penna model (lower plots).

2.7. Ageing and the loss of complexity

When we consider the ageing processes, usually we are thinking about deterioration of physiological functions of organism which leads to the loss of its adaptive possibilities, improper responses to the environmental conditions and eventually to death. Those are only symptoms of ageing; the basic causes of these symptoms are somewhere deeper in the complex living system. Physiologists are aware of the complex nonlinear dynamics of organisms' functions and it is more and more accepted that diseases and ageing could be characterized by the loss of complexity [36, 37]. Even in the simplest organisms, several genes and usually much more gene products interplay with each other [38]. They are connected in the complicated metabolic networks with huge number of different feedback regulatory loops. In fact there is no single function, which is not connected with many other functions of organisms. The interplay between different functions helps the healthy organism to adapt to different environmental conditions. The loss of some connection or improper response to the demand could lead to the disease or death. Such a false response could be connected with genetic defect(s).

For the further extension of the Penna model we have used the version with noise described in the above section;

- the internal "personal" fluctuations can be superimposed on the fluctuations of the environment,
- the sum of both fluctuations describes the health status of organism,
- the fluctuations higher than the assumed limit kill the individual.

In the new version the random Gaussian fluctuations are characteristic for the normal (wild) version of a gene. Defective genes change their random fluctuations for highly correlated signal of the same energy as the previous fluctuations. The idea is biologically justified, since in many examples the highly synchronized behaviors of some activities leads to harmful effects or even deaths [37, 39] (as in case of Huntington disease).

The difference between the new proposed model and the Noisy Penna Model is in behavior of an individual state. In this version we assume that the individual state is a composition of many small fluctuations corresponding to L particular genes, thus

$$P_i(t) = \sum_{j=1}^{L} p_{i,j}(t), \tag{2.3}$$

where

$$p_{i,j}(t) \sim \mathcal{N}(0, \sigma_A^2). \tag{2.4}$$

Defects are reflected in positive correlation between the switched on defective genes. In other words, if genes k and l are defective and switched on then

$$corr(p_{i,k}, p_{i,l}) = \rho.$$

If all $p_{i,j}(t)$ are independent, then $var(P_i(t)) = L\sigma_A^2$, but the correlations between gene functions increase the variation of personal state. The most sharp effect is if the $\rho = 1$. In this case, the variation of L totally correlated signals is equal to $var(P_i(t)) = L^2\sigma_A^2$.

In Fig. 2.17 we have compared the results generated by the proposed model with the results obtained with other models. The main difference is a mortality of newborns. It is much higher than in case of Noisy Penna Model and Standard Penna Model. Since the base variation is equal to composition of $L = 64$ noises, it is high enough to determine the higher mortality.

2.8. Why women live longer than men

Comparing the mortality curves or age structures of men and women of many nations or ethic groups it could be clearly shown that the life expectancy of men is much lower than that of women. Especially, mortality of men at the middle age is almost 50% higher than the mortality of women at the same age. Stauffer blamed for that many natural or social phenomena, like higher somatic mutations rate in the male bodies, stress or altruism of men - they have to fight for peace. But there could be one more reason - male are hemizygous in respect to X chromosome - they possess only one copy of that chromosome in their genome where about 5% of all genes (in the human genome) are localized. Women possess two copies of the X chromosome. It is very important for phenotypic expression of the defective genes located on the X chromosomes. If man has a defective gene on its copy of X chromosome, the phenotypic defective trait would be expressed. The same defective gene on one copy of X chromosome in the woman genome could be complemented by its allele located on the other copy of X chromosome.

Assuming these differences in the genome structure of male and females Schneider *et al.*, and Kurdziel *et al.*, analyzed the age structure of populations with such a sexual dimorphism. They found, that the differences in

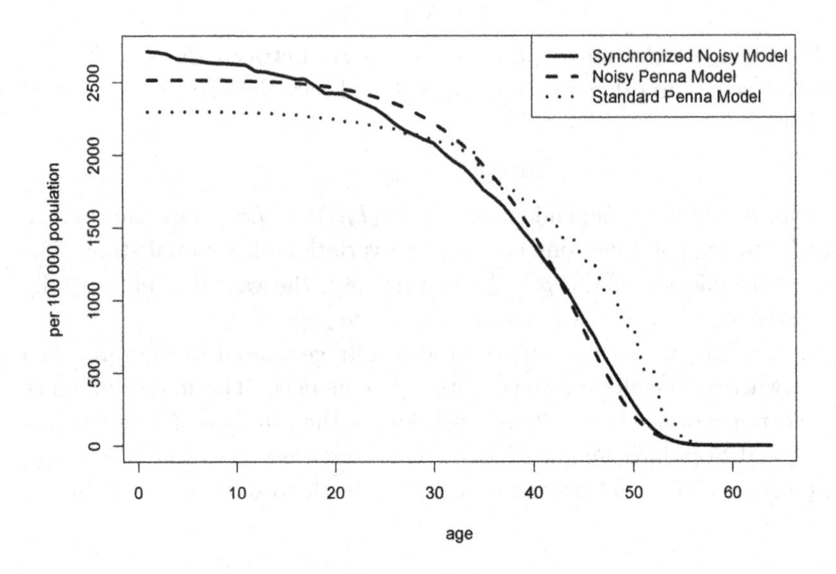

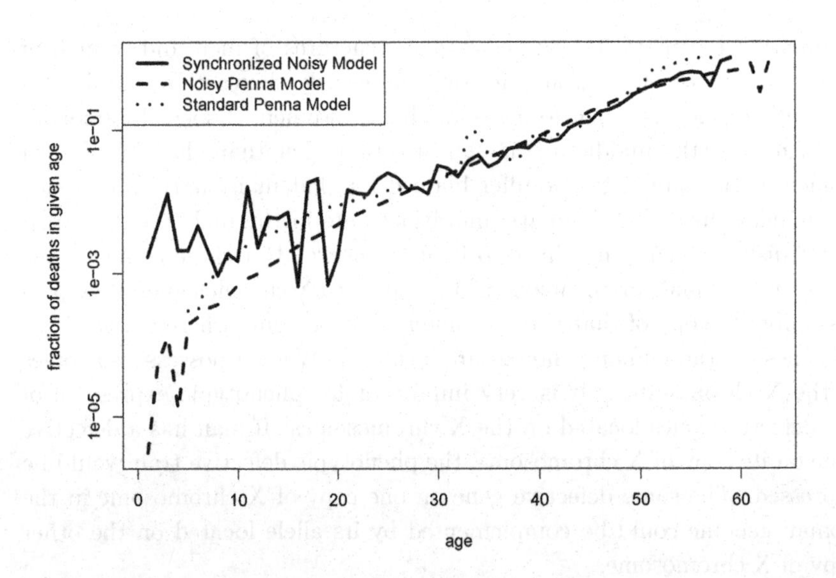

Fig. 2.17. The age distribution (upper plot) and mortality curve (lower plot) for example settings of standard Penna Model (dotted line), the Noisy Penna Model (dashed line) and the proposed synchronized Penna Model (solid line). The exact values of given curves depend on the model parameters.

mortality of both sexes could be explained by the lack of the second copy of X chromosome in the male genomes. Furthermore, the model predicts higher differences in mortality in the middle ages and almost equal mortality for the oldest groups of individuals, just like in human populations. Note, that in this example authors introduced some genes expressed before the birth. The fraction of these genes was big enough to introduce so called zygotic death at the level of 60% corresponding to estimations for humans. It is estimated that a chance for human zygote to survive until birth is of the order of 40%.

Thus, this simple hypothesis of the role of X chromosome has been supported by the simulations. But there is another problem for biologists. Males usually have two sex chromosomes: X and Y. There are some premises that originally these two chromosomes were homologous and in the course of evolution the Y chromosome has shrunk. Why? The Onody group used the Penna model to simulate the evolution of Y chromosome and they noticed that it accumulates mutations with higher probability than other chromosomes. They tried to explain this phenomenon by a specific genetic behavior of the Y chromosome - it does not recombine with counterparts (in fact it has no counterparts) and it is cloned and passes only through the male genomes. Nevertheless, it was found that these properties does not explain the phenomenon of the Y chromosome shrinking. Let's start the evolution of sex chromosomes in the individuals whose genomes are composed of two pairs of chromosomes - one pair of autosomes (autosomes - all chromosomes which are not sex chromosomes) and one pair of sex chromosomes. At the beginning all chromosomes were identical and evolved exactly according to the standard diploid Penna model. The only difference was a marker glued to one copy of sex chromosomes which determined the male sex (the chromosome is called Y chromosome, from now). Thus, if a zygote was formed of two gametes - both transferring the sex chromosomes without that marker (X chromosomes) - it was a female. If a zygote got a sex chromosome with that marker - it was a male. The males could transfer the X chromosome and the Y chromosome to the gametes with an equal probability. Autosomes in all genomes and X chromosomes in the female genomes can recombine.

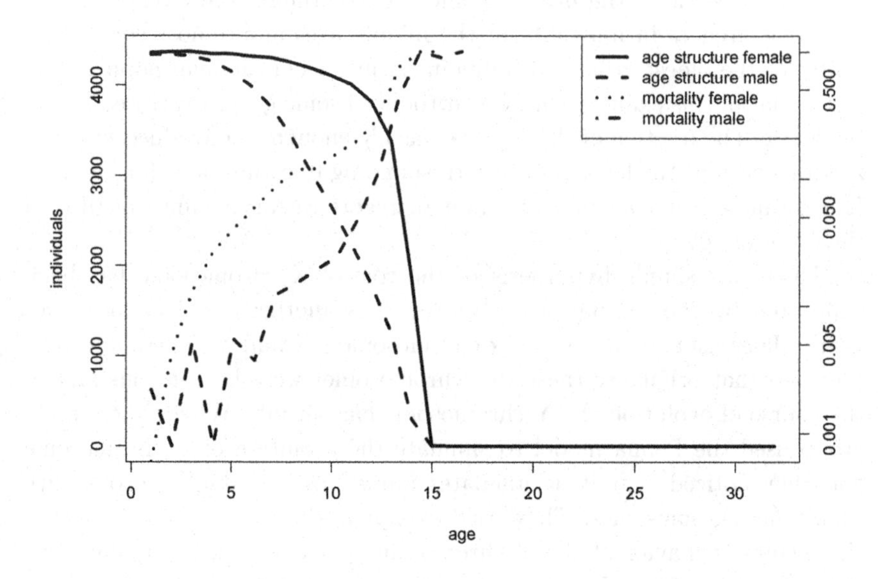

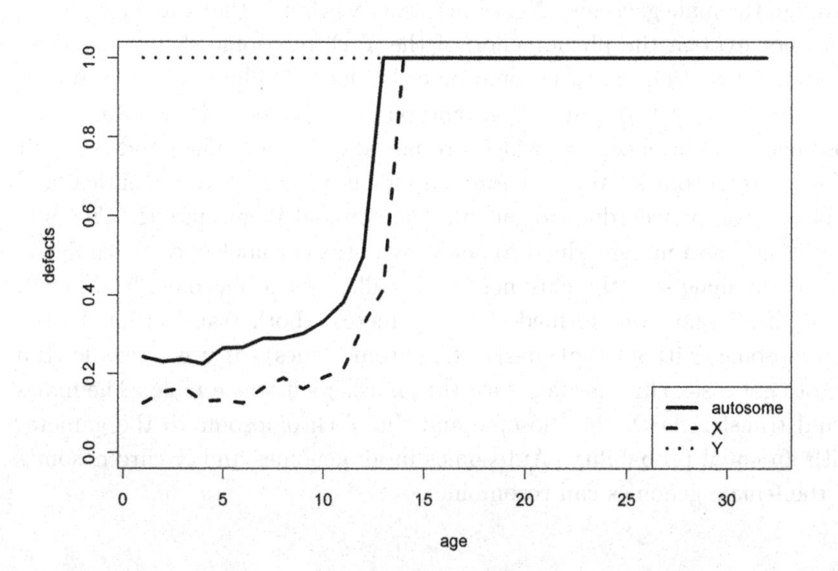

Fig. 2.18. XX/XY system - panmictic population. Age structure and mortality in populations after 20 000 MCs (the upper plots). Note the logarithmic scale (right) for mortality. Lower plots represent fraction of defective alleles in autosomes and sex chromosomes.

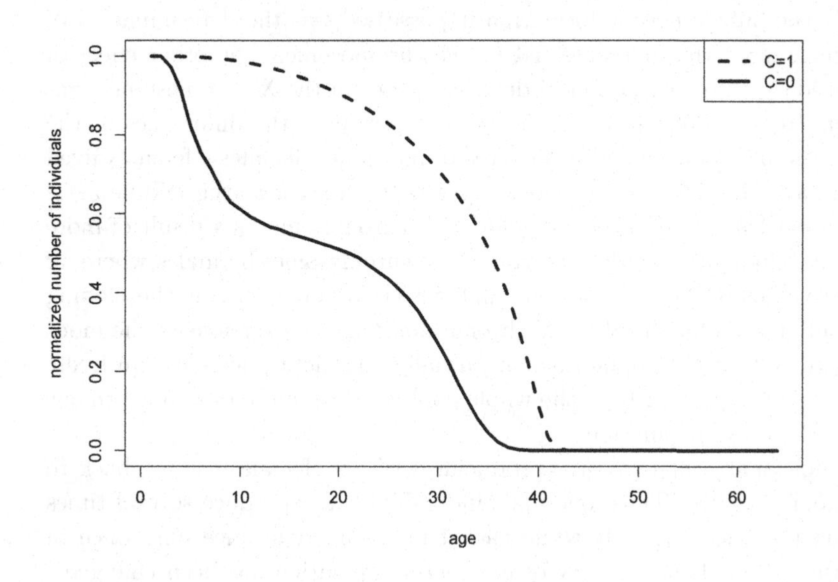

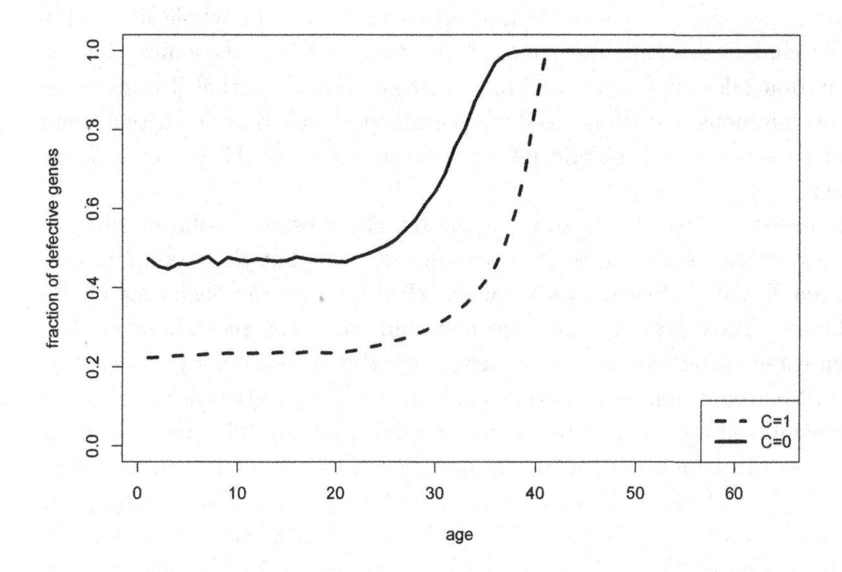

Fig. 2.19. Results for different rates of crossover. Higher crossover rate results in lower fraction of defective genes and higher fraction of individuals in reproduction age.

The result of simulations are shown in Fig. 2.18. After relatively short time of evolution, the Y chromosome accumulated defective genes and eventually the only genetic information it possessed was the determination of the male sex - the marker glued to the chromosome. The other effect of evolution was lower fraction of defective genes in the X chromosome comparing to the autosomes and the most important - the differences in the age structures and mortalities curves of males and females - females lived longer and the difference in mortality was the highest at the middle ages. The lower fraction of defective genes in X chromosomes is a result of more effective elimination of defects from these chromosomes by males where all defective genes play role of dominant defective mutations. Is the cloning the only reason for shrinking Y chromosome? In such versions of the modeling the evolution, populations are usually panmictic - females can freely choose a male partner from the whole pool of males in the reproduction age available in the population.

Additionally, male after reproducing with one female is going back to the pool of males. Thus, one male individual can reproduce several times during the one time unit while each female can reproduce only once in the time step. This property of panmictic population has been changed - male individual after reproduction could not go back to the pool of males in the population, it has to be faithful to its female for whole life. This simple change dramatically changed the fate of Y chromosome. The Y chromosome does not shrink any more. One can conclude that if women are not promiscuous and do not seduce the married men than Y chromosome could preserve its all genetic information and men could live as long as women.

It is still not the whole story of the sex chromosome evolution. In the above versions of the model it has been assumed that the recombination between X and Y chromosomes was switched off from the beginning of the evolution. Biologists are not sure how and when the recombination between these chromosomes was switched off. But if recombination between X and Y chromosome is advantageous then any modification of this process decreasing the crossover rate should be eliminated by selection. To keep the lower recombination rate as an adaptation, one has to assume that it is profitable. Thus, it should be possible to observe self organization of recombination rate during the simulation of the population evolution. At the beginning of simulations, one pair of autosomes, X - X pair and X - Y pair of sex chromosomes recombined with an equal frequency 0.5. The newborns inherited the recombination rate as an average of both parents but this inherited value of recombination rate could be changed in their genomes by +/- 0.01 independently for each pair of chromosomes. During the evolution, the recombination rate between autosomes and between X chromosomes increased and fluctuated around one crossover event per gamete production while the recombination between X and Y chromosome was switched off very fast (it stayed at the low level corresponding to the fluctuations caused by the increment of its changes).

2.8.1. *The role of the crossover rate for the strategies of evolution*

In the standard diploid Penna model recombination between the bitstrings mimics the crossover and usually it is assumed that the frequency of crossover is 1 per bitstrings' pair during the gamete production. There are some other possibilities of the implementation of crossover, very often used in other models. One of them assumes that during the gamete production one allele is randomly drawn from each pair of loci from the bitstrings. This method is not biologically justified because it assumes that all genes are inherited independently - that there are no genetic linkages between genes, which is obviously not true. Here we would like to show how unexpected results can be obtained during modeling the population evolution and how complicated and difficult are biological interpretations of such results.

We have compared the results of two different simulations: with one recombination event between the bitstrings (haplotypes) and without recombination. In the second version, one randomly drawn haplotype after mutation was considered as gamete. The results are shown in Fig. 2.19.

The populations evolving without recombination are smaller and they have very high level of defective genes even in the part of the genomes expressed before the reproduction age. The frequency of defects reaches 0.5. The minimum reproduction age was set for 20, thus, assuming the random distribution of defects at this part of genome and that all mutations are recessive it is easy to calculate the probability of expression the phenotypic trait of a single locus. It equals 0.25. Thus, on average, five defects should be expressed before the reproduction age, substantially decreasing the reproduction potential of population, in fact population should die. Nevertheless, populations survive for very long time and no signs of extinction had been observed. This puzzle will be a subject of the next section which describes the two strategies of genomes' evolution.

Acknowledgements

The work was done in the frame of European programs: COST Action MP0801, FP6 NEST - GIACS and UNESCO Chair of Interdisciplinary Studies, University of Wrocław. Calculations have been carried out in Wrocław Centre for Networking and Supercomputing (http://www.wcss.wroc.pl), grant # 102.

References

[1] P. B. Medawar, *An unsolved problem of biology.* (H.K. Lewis & Co., London, 1952).

[2] D. Harman, Ageing: a theory based on free radical and radiation chemistry, *J. Gerontol* **11**, 298—300, (1956).

[3] D. Harman, The ageing process, *Proc. Natl. Acad. Sci. U.S.A.* **78**, 7124—7128, (1981).

[4] E. Le Bourg, Evolutionary theories of aging: Handle with care, *Gerontol.* **44**, 345—348, (1998).

[5] E. Le Bourg and G. Beugnon, Evolutionary theories of aging: 2. The need not to close the debate, *Gerontol.* **45**, 339—342, (1999).

[6] A. Weismann, *The Germ-Plasm. A Theory of Heredity.* (Charles Scribner's Sons, 1893).

[7] W. D. Hamilton, The moulding of senescence by natural selection, *J. Theor. Biol.* **12**, 12—45, (1966).

[8] B. Charlesworth, *Evolution in Age-Structured Populations.* (Cambridge University Press, London, 1980).

[9] T. B. L. Kirkwood and S. N. Austad, Why do we age? *Nature* **408**, 233—238, (2000).

[10] T. C. Goldsmith, The Evolution of Ageing. How Darwin's Dilemma is Affecting Your Chance for a Longer and Healthier Life, (2004). `http://www.Azinet.com/ageing/Aging_Book.pdf`.

[11] T. J. P. Penna, A bit-string model for biological aging, *J. Stat. Phys.* **78**, 1629—1633, (1995).

[12] S. Moss De Oliveira, P. M. C. De Oliveira, and D. Stauffer, *Evolution, Money, War and Computers.* (Teubner, Stuttgard-Leipzig, 1999).

[13] S. Cebrat and D. Stauffer, Altruism and antagonistic pleiotropy in Penna ageing model, *Theory in Biosciences*, **123**(3), 235—241, (2005).

[14] S. Moss De Oliveira, A. T. Bernardes, and J. S. Sa Martins, Self-organisation of female menopause in populations with child-care and reproductive risk, *Eur. Phys. J. B* **7**, 501—504, (1999).

[15] P. Biecek and S. Cebrat, Immunity in the Noisy Penna Model, *Int. J. Mod. Phys. C* **12**(17), 1823—1829, (2006).

[16] P. Biecek, K. Bońkowska, and S. Cebrat, Relations between organisms

and the environment in the ageing process, *Populations and Evolution* arXiv:0811.0347v1, (2006).
[17] P. Biecek and S. Cebrat, Why Y chromosome is shorter and women live longer? *Eur. Phys. J. B* **65**, 149—153, (2008).
[18] J. S. Sá Martins and S. Cebrat, Random deaths in a computational model for age-structured populations, *Theory in Bioscience* **119**, 156—165, (2000).
[19] K. Bońkowska, P. Biecek, A. Łaszkiewicz. and S. Cebrat, Relationship between the selection pressure and the rate of mutation accumulation, *Banach Center Publ* in press (2008).
[20] K. Kinsella and D. R. Phillips, Global Aging: The Challenge of Success. Population Bulletin, *A publication of the Population Reference Bureau* **60**, 1—40, (2005).
[21] J. F. Fries, Aging, natural death, and the compression of morbidity, *New England Journal of Medicine* **303**, 130—135, (1980).
[22] J. Oeppen and J. W. Vaupel, 2002. Broken Limits to Life Expectancy, *Science* **296**, 1029—1031, (1980).
[23] S. J. Olshansky, B. A. Carnes, and C. Cassel, In search of Methuselah: estimating the upper limits to human longevity, *Science* **250**, 634—640, (1990).
[24] A. I. Yashin, A. S. Begun, S. I. Boïko, S. V. Ukraintseva, and J. Oeppen, The new trends in survival improvement require a revision of traditional gerontological concepts, *Exp. Gerontol.* **37**, 157—167, (2001).
[25] A. Łaszkiewicz, Sz. Szymczak, and S. Cebrat, Prediction of the human life expectancy, *Theory in Biosciences* **122**, 313—320, (2003).
[26] K. Bońkowska, Sz. Szymczak, and S. Cebrat, Microscopic modeling the demographic changes, *Int. J. Mod. Phys. C* **17**(10), 1477—1484, (2006).
[27] A. Łaszkiewicz. *Biological interpretation of the Penna model parameters.* PhD thesis, Wrocław University, Poland (2006).
[28] S. D. Pletcher and C. Neuhauser, Biological Aging - Criteria for Modeling and a New Mechanistic Model, *Int. J. Mod. Phys. C* **11**, 525—546, (2000).
[29] P. Biecek and S. Cebrat, Fluctuations, Environment, Mutations Accumulation and Ageing, *Int. J. Mod. Phys. C* **17**(7), 923—931, (2006).
[30] D. Stauffer, *Annual Reviews of Computational Physics VII.* (World Scientific, Singapore, 2000).
[31] N. S. Gavrilova, L. A. Gavrilov, V. G. Semyonova, and G. N. Evdokushkina, Does Exceptional Human Longevity Come With High Cost of Infertility? Testing the Evolutionary Theories of Aging, *Ann. of the New York Acad. of Scien.* **1019**, 513—517, (2004).
[32] A. Łaszkiewicz, Sz. Szymczak, and S. Cebrat, The Oldest Old and the Population Heterogeneity, *Int. J. Modern Phys. C* **14**(10), 1355—1362, (2003).
[33] J. B. Coe, Y. Mao, and M. E. Cates, Solvable senescence model showing a mortality plateau, *Phys. Rev. Lett.* **89**, 288103, (2002).
[34] V. Schwammle and S. M. De Oliveira, Simulations of a mortality plateau in the sexual Penna model for biological ageing, *Phys. Rev. E* **72**, 031911, (2005).
[35] S. Cebrat, P. Biecek, M. Kula, and K. Bońkowska. 2007. White noise and

synchronization shaping the age structure of the human population, *Proc. SPIE, Noise and Fluctuation in Biological, Biophysical and Biomedical Systems*, 66020O, 1—9, (2005).

[36] L. Lipsitz and A. Goldberger, Loss of „complexity" and ageing: potential application of fractals and chaos theory to senescence, *JAMA* **267**, 1806—1809, (1992).

[37] A. Goldberger, L. Findley, M. Blackburn, and A. Mandell, Nonlinear dynamic of heart failure: Implications of long wavelength cardiopulmonary oscillations, *Am. Heart J.* **107**, 612—615, (1984).

[38] H. Echols, Bacteriophage lambda development: temporal switches and the choice of lysis or lysogeny, *Trends In Genetics* **2**, 26—30, (1986).

[39] L. Lipsitz, Dynamice of stability: The physiologic basis of functional health and frailty, *J. Gerentol.* **57A**, B115—B125, (2002).

[40] J. Schneider, S. Cebrat, and D. Stauffer, Why do women live longer than men? A Monte-Carlo simulation of Penna models with X and Y chromosomes, *Int. J. Mod. Phys. C* **9**, 721—725, (1998).

[41] E. Niewczas, A. Kurdziel, and S. Cebrat, Housekeeping genes and death genes in the Penna ageing model, *Int. J. Mod. Phys. C* **11**, 775—783, (2000).

[42] T. Hassold, Chromosome abnormalities in human reproductive wastage, *Trends Genet.* **2**, 105—110, (1986).

[43] A. Copp, Death before birth: clues from gene knockouts and mutations, *Trends Genet.* **11**, 87—93, (1995).

[44] M. P. Lobo and R. N. Onody, Degeneration of the Y chromosome In evolutionary aging models, *Eur. Phys. J. B* **45**, 533—537, (2005).

[45] A. Łaszkiewicz, E. Niewczas, Sz. Szymczak, A. Kurdziel, and S. Cebrat, Higher mortality of the youngest organisms predicted by the Penna aging model, *Int. J. Modern Phys. C* **13**, 967—973, (2002).

[46] A. Łaszkiewicz, Sz. Szymczak, and S. Cebrat, Speciation effect in the Penna aging model, *Int. J. Modern Phys. C* **14**, 765—774, (2003).

[47] B. Gompertz, On the nature of the function expressive of the law of human mortality, and on a new mode of determining the value of life contingencies, *Phil. Trans. Roy. Soc.* **115**, 513—585, (1825).

Chapter 3

Darwinian purifying selection versus complementing strategy in Monte-Carlo simulations

Wojciech Waga, Marta Zawierta, Jakub Kowalski, and Stanisław Cebrat

Department of Genomics, Faculty of Biotechnology,
University of Wrocław,
ul. Przybyszewskiego 63/77, 51-148 Wrocław, Poland.
cebrat@smorfland.uni.wroc.pl.

Intragenomic recombination (crossover) is a very important evolutionary mechanism. The crossover events are not evenly distributed along the natural chromosomes. Monte-Carlo simulations revealed that frequency of recombinations decides about the strategy of chromosomes' and genomes' evolution. In large panmictic populations, under high recombination rate the Darwinian purifying selection operates keeping the fraction of defective genes at the relatively low level. In small populations and under low recombination rate the strategy of complementing haplotypes seems to be more advantageous. Switching between the two strategies has a character of phase transition - it depends on inbreeding coefficient and crossover rate. The critical recombination rate depends also on the size of chromosome. It is also possible, that in one genome some chromosomes could be under complementing while some other under purifying selection. Such situation stabilizes genome evolution and reproduction strategy. It seems that this phenomenon can be responsible for the positive correlation between kinship and fecundity, recently found in the Icelander population. When large population is forced to enter the complementing strategy, the phenomenon of sympatric speciation is observed.

Contents

3.1. Introduction

It is difficult to define the term "species" in a way that applies to all naturally occurring populations. Usually a species is defined as "all individual organisms of a natural population which interbreed at maturity in the wild and whose interbreeding produces fertile offspring". Note words: natural populations, the wild and fertile offspring. In unnatural conditions, a lot of barriers can be broken - in captivity or in laboratories, interbreeding of individuals belonging to different species could be successful. On the other hand, mule is not a species, because mule mating with mule never gives an offspring. Another definition of species is "a pool of genes which could be freely exchanged between individuals belonging to the population". The last definition is particularly plausible for geneticists. Species could be extinct thus, species have to appear. Appearing of new species is called speciation. There are two distinct ways of speciation: allopatric and sympatric speciation. There are no problems with allopatric speciation when populations of one species are divided by physical, geographical or biological barriers, and they even physically cannot interbreed for a long time. They eventually form two (or more) different species which cannot produce fertile offspring by interbreeding - such a way of emerging of new species is called allopatric speciation.

Sympatric speciation is a quite different phenomenon - it is an emergence of a new species inside the older one on the same territory without any physical or geographical barriers. Since Ernst Mayr [1], who was rather sceptical about the possibility of sympatric speciation, this phenomenon is still debatable [2], though some well documented empirical data that support it, already exist [3], [4]. Also some theoretical models that address the problems showed the possibility of speciation in sympatry [5], [6]. In this section we are going to show in computer modeling that sympatric speciation could be the intrinsic property of the evolving populations.

3.2. Mutations, frequency of defective alleles in the populations and complementation

Mutation is a change in a sequence of nucleotides in a genome, it can transform the functional gene into its defective unfunctional form (see Chap. 2). Many experimental and theoretical analyses suggest that the level of mutational pressure is of the order of one mutation per genome per generation, independently of the genome size [10], [9]. It is even possible, that there is a phase transition between the "order phase" where the mutational pressure is low enough to enable the population survival and "disorder phase" where the mutations happen too frequently to be effectively eliminated by selection [7]. According to neo-Darwinism, mutations are random thus, they could happen in any gene and it is selection which eliminates some of them with higher efficiency than others. If homozygous state (when individual has both defective alleles in the locus) is lethal then, in the Mendelian populations (infinite in size), where we assume random mutations, random crossovers with high frequency and no mating preferences we should expect for lethal mutations:

$$p_d = f^2, \tag{3.1}$$

where:

p_d - probability of genetic death caused by alleles of single locus,
f - fraction of defective alleles in this locus in the whole genetic pool of interbreeding population.

Thus, probability of death corresponds to probability of meeting two defective alleles at the same position in one diploid genome. It depends on the frequency of defects at these positions in all bitstrings (haplotypes) in the population (genetic pool). Since selection eliminates defective alleles from the genetic pool by the genetic death, their frequency would decrease with time if there are no new mutations. Nevertheless, mutational pressure introduces new defective alleles into the genetic pool and, after long time of evolution equilibrium should be established with a level of defective genes corresponding to the force of selection and mutational pressures. The phenomenon of elimination the defective genes from the genetic pool of population is called the purifying selection. One can expect that if all genes have the same selection value and the same probability of mutations, the frequency of defects in the genetic pool should be at the same, rather low level, in all loci.

In the situation described in the above paragraph the effect of defective gene can be invisible - phenotype of the organism is normal if only one copy of the two corresponding alleles is defective and the defect is recessive. It is said that the mutation or defective gene is complemented by the normal, functional one. In our model we assume that all defective genes are complemented by their normal counterparts. That is why it is imaginable at the extreme situation that at each locus of the diploid genome composed of two bitstrings one bit is set to 0 while the other one is set to 1. Such a structure of genomes is called the complementing haplotypes. Of course it is highly improbable in the model described above for the Mendelian populations. We have described in the previous chapter, that nonrandom distribution of genes on chromosomes generates some preferences in elimination of defective genes (Chap. 1). In the Penna model [8], the differences in the effectiveness of elimination the defective alleles depend on the period of life when they are expressed.

What can we find in the natural populations? There are many known examples of very high frequencies of recessive defective genes in the human population *i.e.* cystic fibrosis, sickle cell anemia or Tay-Sachs disorder. In the Caucasian race, one in every 25 persons is a carrier of a defect responsible for cystic fibrosis, among African-Americans one in 12 is a carrier of S hemoglobin gene responsible for sickle cell anemia and one in every 27 of Ashkenazi American Jews is a carrier of defective Hex-A gene responsible for Tay-Sachs disorder (carrier is an individual with one defective, recessive allele and a second one in the corresponding locus - correct). All these examples have one common property - the incidence of the defective alleles is not universal in the human populations, they have rather endemic character or are characteristic for some ethnic groups. What are the reasons of such a high frequency and such an uneven distribution of these defects? The best characterized is a sickle cell anemia (see [21] for review). The point mutation in a gene coding for hemoglobin results in a replacement of one, proper amino acid by a wrong one and the "wrong" hemoglobin is called S hemoglobin. See the Chap. 1 for an explanation on how single mutations in DNA cause the amino acid substitutions in proteins. People possessing one gene for S hemoglobin and the other one for normal hemoglobin are carriers and it is said they have sickle cell trait. These people are normal in almost all respects and rarely develop problems related to their genetic conditions thus, we can state that the mutation is recessive. Could we say that selection does not discriminate the carriers? Sickle cell trait provides a survival advantage over people with normal hemoglobin in regions where malaria is endemic. In fact this trait provides neither absolute protection nor resistance to the disease. People, and particularly children, infected with *Plasmodium falciparum* (which is a parasite responsible for malaria) are more likely to survive the acute illness if they have S hemoglobin gene - have a sickle cell trait. Each year, malaria attacks about 400 million people, two to three million of whom succumb to the illness. From our point of view, each year, natural selection (malaria) tests 400 million people discriminating the normal ones and favoring the carriers of S hemoglobin genes.

The other example is cystic fibrosis - carriers suppose to be more resistant to the infections of the alimentary tract. In a case of Tay Sachs disease interpretation is not so simple, some people suggest that carriers are more intelligent [20], [19]. Even if we assume it is true - does intelligence increase the survival probability and fertility?

Taking into account the above examples it is possible to model the evolution of population where selection favors heterozygotes (carriers) which have one correct allele and one defective while defective homozygotes (both alleles in the corresponding loci are defective) are lethal.

3.3. Assumptions of the model

Our virtual population is composed of N individuals. Each individual is represented by its diploid genome - two bitstrings L bits long (haplotypes). Bits at the same positions (loci) in both bitstrings correspond to alleles. Bit set to 0 corresponds to the wild allele (correct one) while bit set to 1 corresponds to the defective allele. All defective alleles are recessive, which means that both alleles at the corresponding positions should be defective to determine the defective phenotypic character. Each phenotypic defect is lethal - an individual with such a trait has to die. The whole genomes of newborns are checked at birth. If at any position of bitstrings both alleles are defective, the individual is eliminated otherwise it stays at the population. One can notice that in this model there are no ageing and all positions in the bitstrings seems to be equivalent. After each Monte-Carlo step (MCs) the declared fraction of a population is randomly killed (usually 5%). To fill up the gap, in the next MCs randomly chosen individuals can reproduce. To produce an offspring, both chosen individuals produce the gamete: they copy their two haplotypes, introduce mutations into each copy of haplotype with a probability M and perform a crossover between the pair of new haplotypes with a probability C. Mutation replaces 0 bit by 1, if the bit chosen for mutation is already 1 it stays 1 - there are no reversions. Two gametes, each produced by one of the two parental individuals are joined and form a genome of the newborn which is immediately checked for its genetic status. The MCs is completed when the population reaches the declared size of N individuals. The main difference between this model and the Penna model is in the way how the genome is checked - in the Penna model it is checked chronologically, in this model all loci are checked at birth.

3.4. Positive selection for heterozygosity

To show how the frequency of defective alleles could be affected by positive selection for heterozygous loci an additional condition has been introduced: the chance for reproduction of the organism is proportional to (h+1) where h represents the number of heterozygous loci in the genome (in a heterozygous loci alleles have different values - one allele is set to 0 and the second one is set to 1). Simulations have been performed for two different strategies of reproduction - cloning, asexual reproduction without any exchange of information between parents and, sexual reproduction as described in the above section but without recombination between haplotypes during the gamete production - (C=0). Notice that there are two possible modes of sexual reproduction for C=0. In the first one, bitstrings are enumerated - the first one and the second one. The first parental genome is a donor of a copy of its first bitstring and the second parent is a donor of a copy of its second bitstring. In this situation there are two separate sets of bitstrings. In the second mode parents are donors of randomly chosen copies of their bitstrings.

Simulations were performed for parameters: L=100, N=1000, M=1. The results of simulations are shown in Fig. 3.1. In a case of cloning, after relatively short time of simulation, all genomes became heterozygous in all loci. The results seem to be trivial because all homozygous defective loci eliminate the genomes (individuals) from the population while heterozygous loci increase the probability of their reproduction. The results of the sexual reproduction without crossover are not so obvious. Nevertheless, after longer simulations (200 000 MCs) all genomes become heterozygous in all loci which means that the frequency of defective alleles in the genetic pool is 0.5 for all loci. If we assume that the mutations are distributed randomly

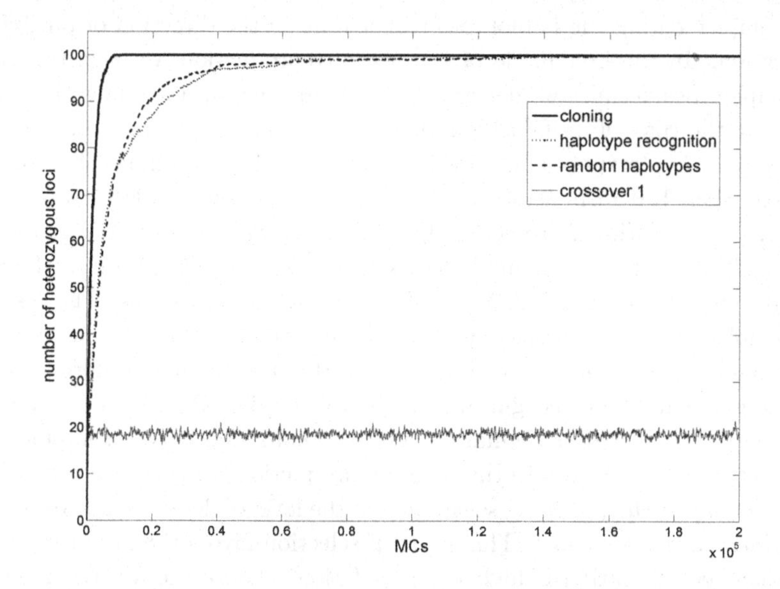

Fig. 3.1. Evolution of genomes under different strategies of reproduction with a positive selection for heterozygous loci. Recombination between haplotypes prevents accumulation of defective genes. Cloning or production of gametes without recombination leads to the high level (100%) of heterozygous loci.

along the haplotypes in the whole genetic pool, then the probability of forming the killing configuration in a single locus $(1\|1)$ is 0.25 while surviving - 0.75. Thus, probability of forming the surviving offspring should be of the order of $0.75^{100} \approx 10^{-13}$. Populations would have no chance to survive under such conditions. Nevertheless, our virtual populations are not going to die. We have checked the Hamming distance (H) between all haplotypes in the populations. Hamming distance is a sum of differences between bits at the same position of bitstring counted for the whole length of the compared pair of bitstrings - it corresponds to the number of heterozygous loci in the genome in our case. After long simulations the Hamming distance for any combination of haplotypes equals 100 or 0 which means that in the whole populations only two types of haplotypes occur. Let us call them $T1$ and $T2$. $T1$ haplotype perfectly complements the $T2$ one thus, $T1||T2$ and $T2||T1$ survive. But $T1|||T1$ and $T2|||T2$ combinations are lethal. Nevertheless, the probability of forming the surviving configuration of haplotypes in such a population is 0.5 instead of 10^{-13} in case of random distribution of defects on haplotypes. That is why population evolves in the direction of complementing the haplotypes rather than in the direction of purifying selection. In the first mode of the sexual reproduction we have two sets of haplotypes because we declared that the first parent introduces into the gamete the copy of its first haplotype and the second parent - the copy of its second haplotype. Thus, we have introduced some kind of haplotype recognition. Is it responsible for choosing the complementation strategy by evolving populations? To check this possibility we have not declared such recognition in the second mode of sexual version. In that strategy, the gametes were produced of randomly drawn parental haplotypes. The results resembled the previous ones - populations evolved toward the complementing haplotypes, though the dynamics of this evolution is different than in case of haplotype recognition, which will be described later in details. In the further series of simulations the recombination between haplotypes (C=1) was introduced into the sexual reproduction. In this case the populations choose the purifying selection and the level of defective alleles in the haplotypes was kept low. Thus, even if selection favors the heterozygosity - when recombination is high the populations choose the purifying selection. One can argue that under the reproduction without recombination between haplotypes the complementation is driven by selection favoring the heterozygosity.

3.5. Is the complementation strategy possible without an advantage of heterozygosity?

The results presented at the end of the last section of the chapter describing the Penna model Chap. 1 suggest that complementation strategy is possible even if there is no declared advantage of heterozygous loci. To show this phenomenon in the simplified model without age structure, we have repeated the simulations described in the above section but without declaring any preference for heterozygosity. All simulations with cloning or without recombination during the gamete production gave the populations with complementing haplotypes. The only simulation showing the Darwinian purifying selection was that one with recombination between haplotypes before production of gametes. The distribution of defective alleles along the haplotypes after the evolution with and without the crossover is shown in Fig. 3.2. The results clearly suggest that there are two possible strategies of genomes' evolution. The first one based on Darwinian selection eliminating defective phenotypes and the second one which allows higher

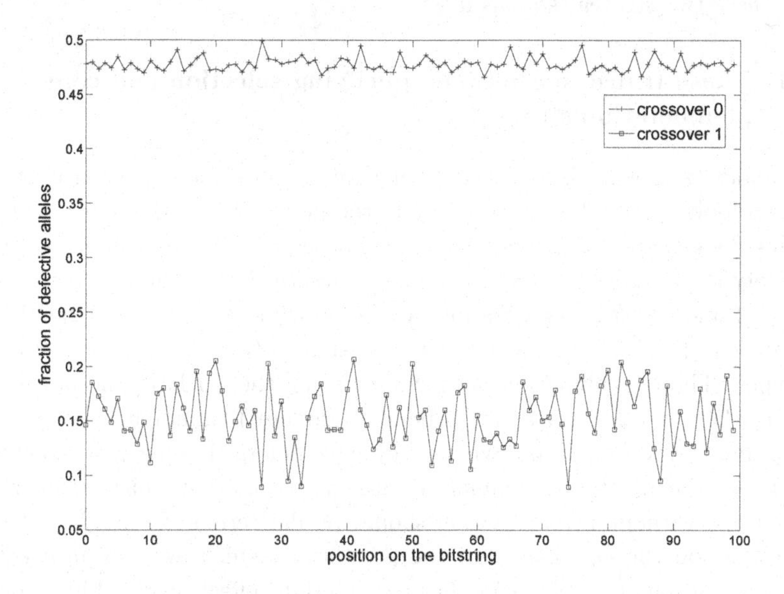

Fig. 3.2. Distribution of defective genes along the (bitstrings) haplotypes in populations evolving without recombination or with 1 recombination per gamete production.

accumulation of defective alleles but it requires the complementation of defective alleles by the wild forms of the genes. The second strategy leads to appearing of specific and unique distribution of defective genes along the haplotypes. Thus, two problems arise, now:

- the first one is connected with that "uniqueness" - each simulation produces specific sets of complementary haplotypes - specific means that distribution of defective alleles along the haplotypes is unique and specific for each independent simulation. The probability that haplotypes from independently simulated populations would complement is equal to 0.5^L, where L is the number of loci in the haplotype. Geneticists would say that crossbreeding between two such populations gives no surviving offspring or unfertile offspring - these two populations suppose to be different species,
- the second problem arises with the efficiency of recombination. If the crossover frequency of the order of 1 per one pair of haplotypes is enough to push the population toward the purifying selection and crossover 0 favors the complementing strategy - then, how the transition between

these two strategies looks like?

3.6. Phase transition between purifying selection and complementation strategy

To understand what is going on during simulations under our parameters, take a look at Fig. 3.3. If we simulate the population without recombination - the genomes of individuals are composed of two fully complementing haplotypes $T1$ and $T2$, according to the description in the above section. $T1|||T1$ and $T2|||T2$ configurations are lethal while $T1|||T2$ and $T2|||T1$ are surviving. Imagine one recombination event between the $T1$ and $T2$ haplotypes. The recombination products, let's call them $T1xT2$, do not fit to $T1$ neither to $T2$. Thus, such products (gametes) cannot form surviving offspring in the population where they have emerged. One may conclude that introducing the recombination into the population which was earlier simulated without recombination should be deleterious for it. It is true. Introduction the crossover into the population which evolved until equilibrium without recombination has a deleterious effect on it. What would happen if the populations are evolving from the beginning at the intermedi-

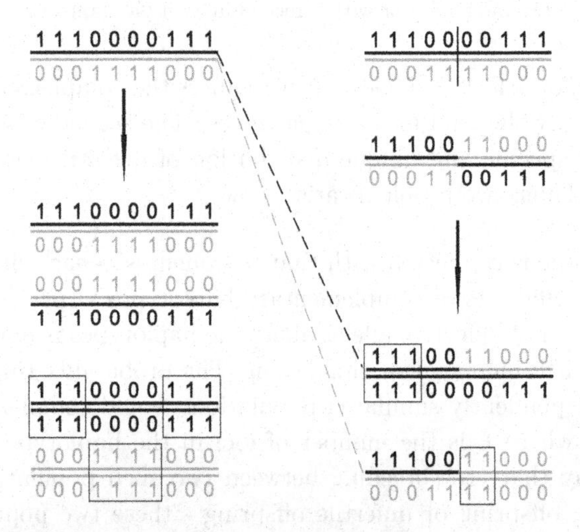

Fig. 3.3. Why recombination prevents the complementation strategy?. If in the population evolving without recombination (left part) only two types of haplotypes occur (bold and light) the probability of forming the surviving newborn equals 0.5. But if in one individual during the gamete production recombination happens, then the possibility of forming a surviving newborn is negligible. The gamete has to find a proper gamete after recombination at the same point.

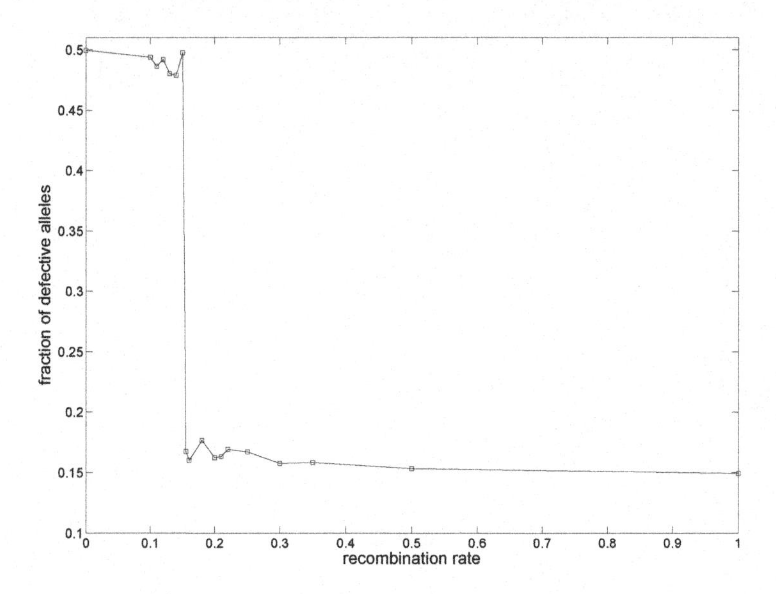

Fig. 3.4. Effect of different recombination rates on the frequency of defective alleles (and the strategy of genome evolution). Parameters of simulations: L=100, M=1, N=1000.

ate crossover rate? We have performed a series of simulations with different crossover rates. All other parameters were constant and their values were like in the standard model. The results are shown in Fig. 3.4. There is a very sharp transition from complementing strategy, where the fraction of defective alleles is close to 0.5, (or fraction of heterozygous loci is close to 1), to the purifying selection where the fraction of defective loci is much lower. The point of transition is around the recombination rate 0.14. To understand better what is going on at this critical point we have analysed the relation between the fertility of individuals and crossover rate (Fig. 3.5). The criterion was the zygotic death - how many trials of forming the diploid genome of a zygote should be done to succeed in producing one surviving newborn. At the transition point (around recombination rate 0.14) the number of trials is the highest which means that the fertility is the lowest - populations should avoid such conditions and "escape" to the regions where the fertility is higher - deeper in purifying strategy or in complementing strategy. The problem is "how population can escape from those fatal conditions". Intragenomic recombination rate is evolutionary established and

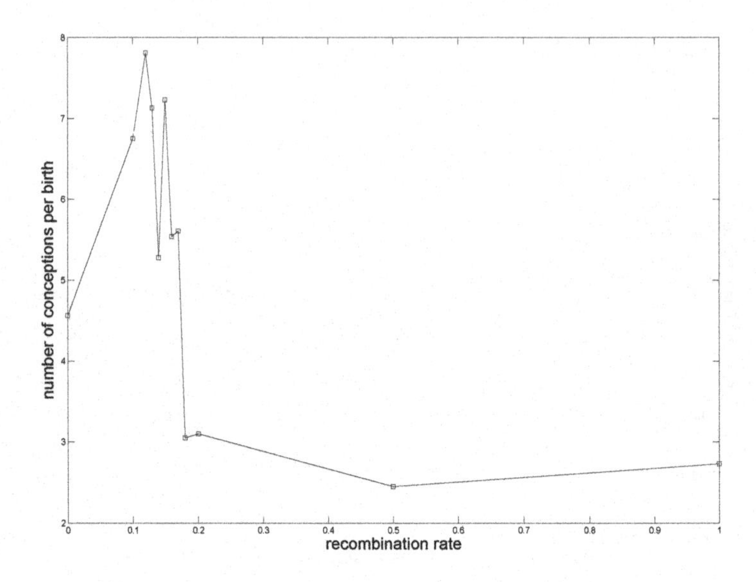

Fig. 3.5. The relation between the recombination rate and average number of conceptions per one successful birth.

it cannot be changed ad hoc. If we think how the complementing strategy is possible at all, we have to take into account the probability of meeting two complementary haplotypes, the most probably of the same origin, from the same ancestor, before they (haplotypes) mutated and recombined with other haplotypes. Before mathematicians will solve this problem formally and exactly, geneticists can say that it should depend on the size of populations. Let's try to look for a transition point in simulations performed with populations of different sizes. Results are shown in Fig. 3.6. Note, both axes are in a logarithmic scale which means that we have found a power law relation between the critical recombination rate and the size of population. The results are very interesting because of that power law. Another interesting biological question which arises is: does the value of recombination frequency at the transition point depend on the bitstring length? Figure 3.7 shows the relations between the transition points, bitstring lengths and populations' sizes. For bitstrings composed of 2048 bits, the critical recombination rate is around two crossover events per bitstring for effective population size in between 100 - 300. The largest bitstrings in our simulations correspond to the length of the largest human chromosomes

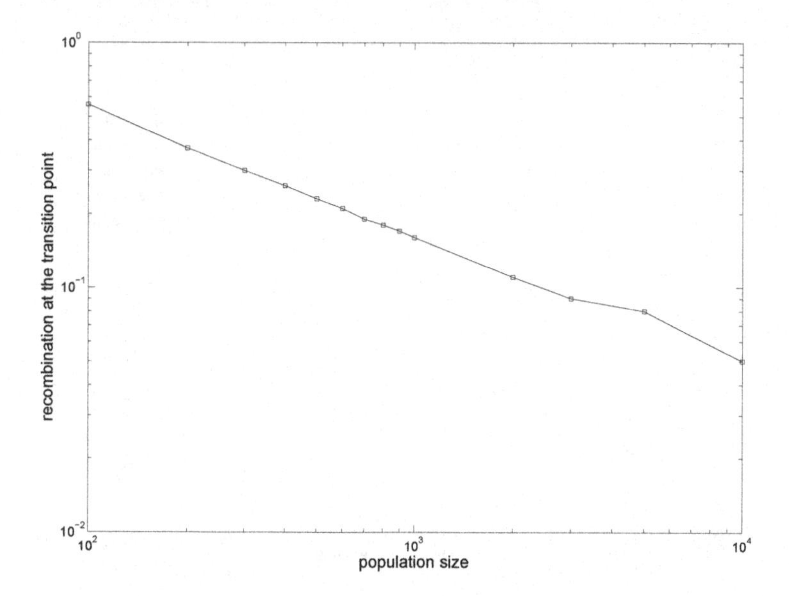

Fig. 3.6. The relation between the frequency of recombination at the transition point (see Fig. 3.4 and Fig. 3.5) and the populations' sizes. Notice the logarithmic scales of both axes.

containing about 2000 genes and in fact there are two crossovers during the meiosis (gamete production) between these chromosomes, on average. Does the Nature operate close to the phase transition? Nevertheless, physicists can't say that it is a phase transition because the transition point in our case depends on the population size. But there is a trap - in all our simulations we have modeled the panmictic populations. In panmictic population an individual looks for partner for reproduction in the whole population. It is not realistic in Nature. In natural populations partners are found in close vicinity, rather. The distance or the size of groups of individuals where the individuals find their sexual partners should be considered as a real effective population size. This effective size of population determines other important genetic parameter characterizing the population - the inbreeding coefficient which is a measure of genetic relations between members of populations and more precisely between the sexual partners.

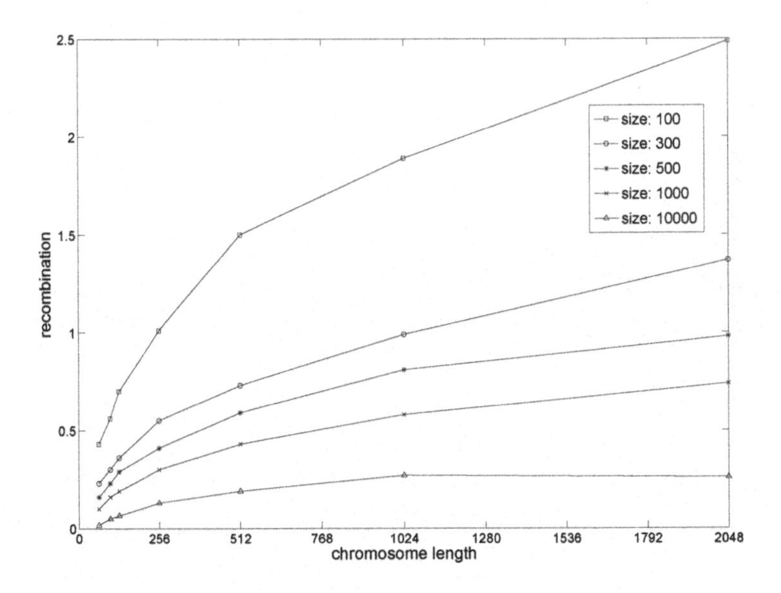

Fig. 3.7. The critical recombination rates for chromosomes of different lengths and different sizes of evolving populations. Above the lines chromosomes are under purifying selection below the line are under complementation strategy.

3.7. Simulations on a square lattice

To analyze the effect of inbreeding and effective population size on the switching between the two different strategies of evolution we have modified the model. Simulations were performed on a square lattice. Each individual occupies one field. Females are looking for their partners at a distance not larger than P and newborns are put in a free square found at a distance not larger than B from their mother. If there is no partner in the range of P or no free place at the range of B - an offspring is not born. Both distances - P and B - determine the effective size of the population (size of subpopulation where individual can look for reproduction partner), though there are some virtual borders limiting the crossbreeding but subpopulations overlaps and in fact there are no physical or geographical borders. To avoid the edge effects at the edges of the lattice (the inbreeding conditions are different because individuals have neighbors only at one side) we have wrapped the lattice on a torus. All other parameters of modelling were the same as in the previous simulations of panmictic populations. In Fig. 3.8

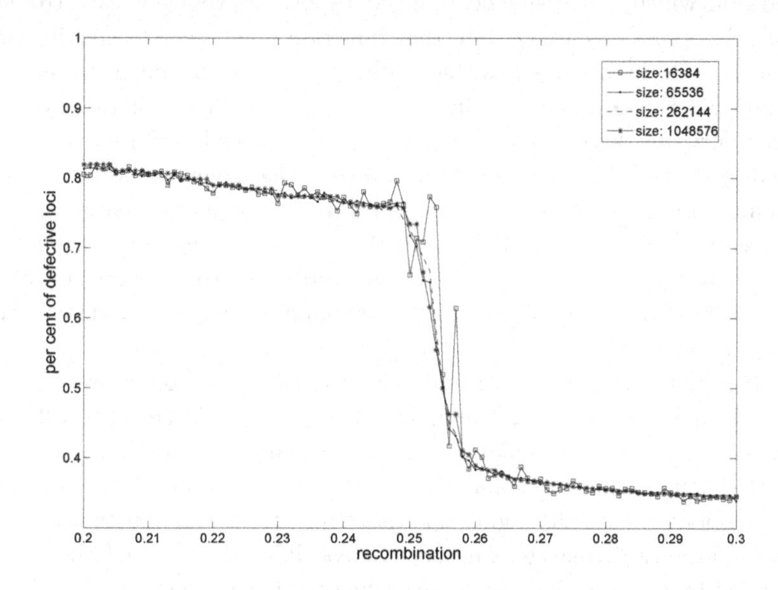

Fig. 3.8. Phase transition between two strategies of the genomes' evolution - purification selection above crossover frequency 0.260 and haplotypes complementation below 0.250. Simulations were performed on square lattices (128×128 - 1024×1024) wrapped on toruses. The other parameters: P=B=2, L=64. Notice that transition point does not depend on the population size.

we have presented the results of simulations for four different sizes of populations (lattices: 128×128, 256×256, 512×512 and 1024×1024). For the fixed P = B = 2 parameters, the value of critical recombination frequency does not depend on the population size - plots for all populations overlap. Note that the range of recombination frequency inside which the population changes dramatically their strategy of evolution is very narrow - between 0.250 and 0.259. The results suggest that the effect can be classified as a phase transition because the value of recombination rate at the transition point does not depend on the population size. Populations choose the more advantageous strategy on the basis of interplay between the intragenomic recombination rate and inbreeding coefficient. The intragenomic recombination rate is an intrinsic, biological property of a genome. It is a result of a long evolution, rather and it can not be changed on a spot. On the other hand, inbreeding coefficient (the probability that the two alleles for any gene are identical by descent) could be considered as an environmental condition which can easily be changed by overcrowding or catastrophes. Thus, if populations are close to the transition point some interesting phenomena could be observed as the results of influence of the environmental conditions. For example, imagine that large panmictic population evolving under purifying selection suddenly shrinks, becomes less dense. Then, inbreeding in it could increase to such a degree that complementing strategy becomes more advantageous and, as a result, we should observe speciation. To observe such effects we have to further modify our model on the lattice in order to recognize different species or genotypes. To accomplish that we colored the square occupied by an individual according to the distribution of the defective alleles on its haplotypes.

Since our computer system can choose one of 2^{24} colors, we ascribed one color to each of $1 - 2^{24}$ numbers then we cut off the central 24 bits long substring of each haplotype of a given individual (*i.e.* bits 21 to 44 in case of simulating the populations with L=64), converted substrings for the numbers, chosen the higher one (note that the two numbers describing two haplotypes of one genome are always different), colored the square occupied by this very individual according to that number. The black and white edition of the paper can not show the full strength of the method - take a look at our web page, see [23].

3.7.1. *Sympatric speciation*

As it has been mentioned in the Introduction, sympatric speciation is an emergence of the new species inside an old one without any barriers - physical or biological. This phenomenon is still debatable because it is difficult to imagine the evolution of genetic systems which could lead to such a speciation. There are many ways to show the sympatric speciation in our model. Let's try the simplest one and start from one pair - Adam and Eve - two parents with perfect genomes (all bits set to 0) in the middle of lattice 1024x1024. The whole simulation is performed according to the standard model, with P and B distances declared. Population starts to expand on the lattice. The most critical values of simulations are P and B parameters - determining the inbreeding (environmental conditions) and, crossover rate determining the intragenomic conditions. When the P and B are low which correspond to small panmictic populations and high inbreeding, even for relatively high crossover rates the populations choose the complementing strategy (for P=B=2 critical recombination equals 0.25, see the above section). When simulation starts from the center of the lattice, after a short time, when genomes accumulate enough mutations to choose the strategy, the whole population is much larger than the effective populations and in fact individuals at the edges of expanding population have no chance to interbreed freely with all individuals, particularly with those at the opposite site of the territory. They evolve almost independently. Thus, the distribution of defective genes along their haplotypes is also different and they cannot form surviving offspring - they are just different species. You can observe the radiating territories of different colors. The same colors correspond to the same haplotypes' configurations, which means that individuals belong to the same species, see [23]. There is another possibility of observing the speciation in this system, much more beautiful and more convincing. Imagine very large panmictic population on lattice. It could evolve under relatively high P and B parameters and low recombination rate. Suddenly, the P and B parameters decrease. It could happen as a result of catastrophe when both, the density of populations and the distances covered by individuals decrease. In such conditions many of focuses of speciation arise and the spectacular speciation could be observed (like during the Cambrian era), see [23].

3.7.2. *Some snapshots of expanding populations*

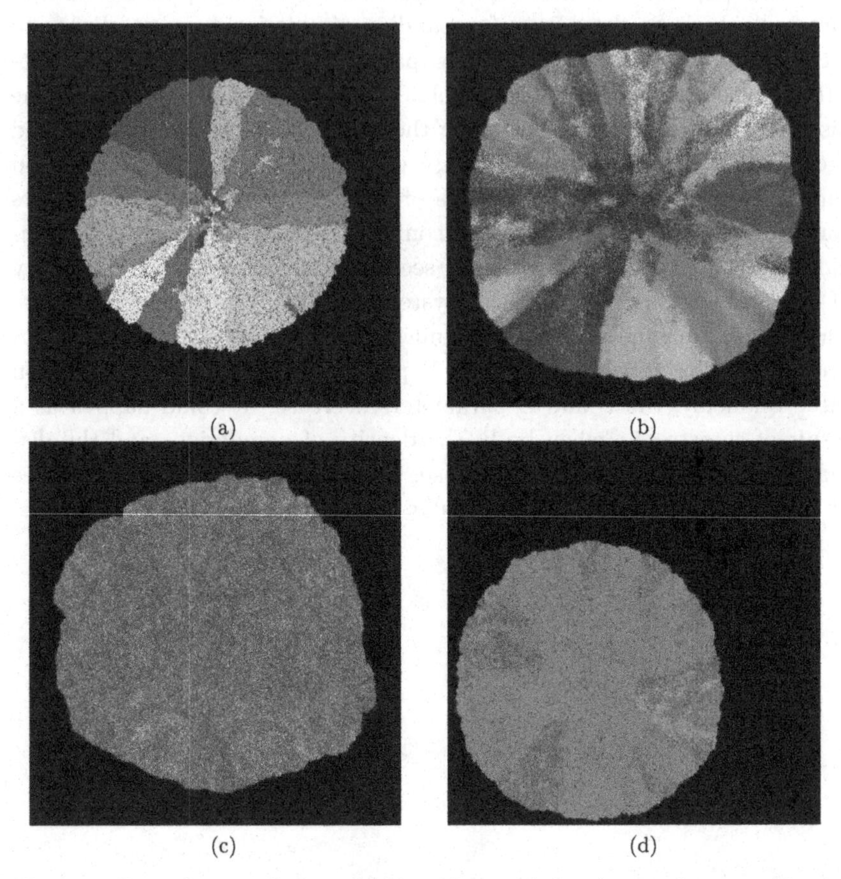

Fig. 3.9. Expanding populations. (a) Very high speciation due to a low recombination (C=0.1) and high inbred (P=B=2). (b) Lower inbred (P=5) and less explicit speciation. (c) No speciation, because of a high recombination (C=0.3) and high inbred (P=B=5) (d) Back-speciation from existing species.

3.7.3. *Expansion rate and crossover frequency*

Let's start again with one perfect pair of parents in the middle of the lattice 1000x1000, P=B=5, crossover rate 0.1. It is obvious, that the expansion rate depends directly on P and B parameters. But at the edges it could be important how the role of the parameters deciding about inbreeding is

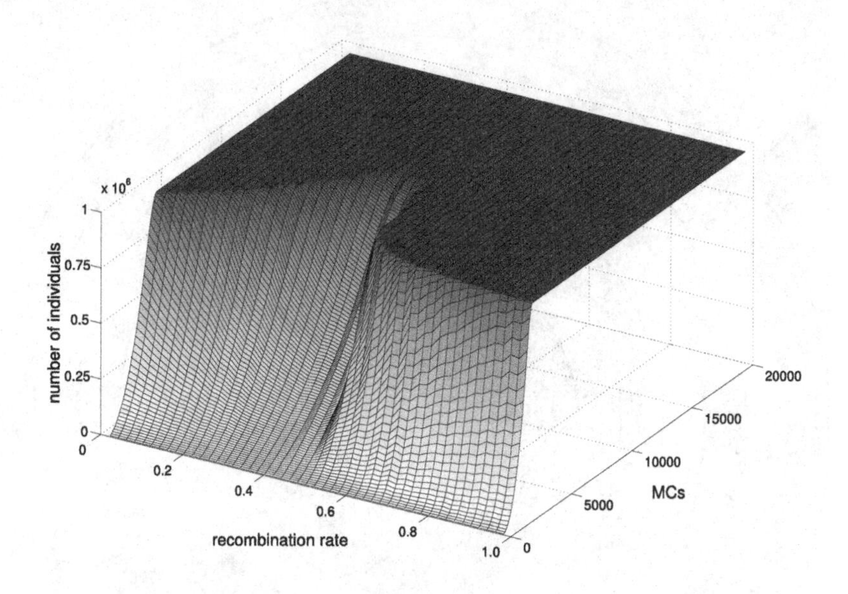

Fig. 3.10. Relation between the expansion rate of population and intragenomic recombination rate.

influenced by crossover rate. One can notice that there are quite different conditions of evolution at the edges of expanding populations and inside it. At the edges, pioneers live under much higher inbreeding than inside. The question is -how the expanding rate depends on the intragenomic recombination rate under constant P and B parameters. Figure 3.10 shows the complicated, nonlinear relations between the expansion rate and crossover rate with P=B=2. The minimum of expansion is observed for crossover about 0.4. Lower or higher recombination rate increases the expansion. One can conclude, that if the population is forced to evolve longer under such conditions the recombination rate should self-organize to adapt to these conditions. It was observed in other simulations with the Penna model [11].

3.7.4. *Geographical distribution of defective genes*

The fractions at the edges of expanding populations evolve under different condition than inside. But they are not isolated and they should influence the interior. Moreover, in our model the old individuals don't move, they

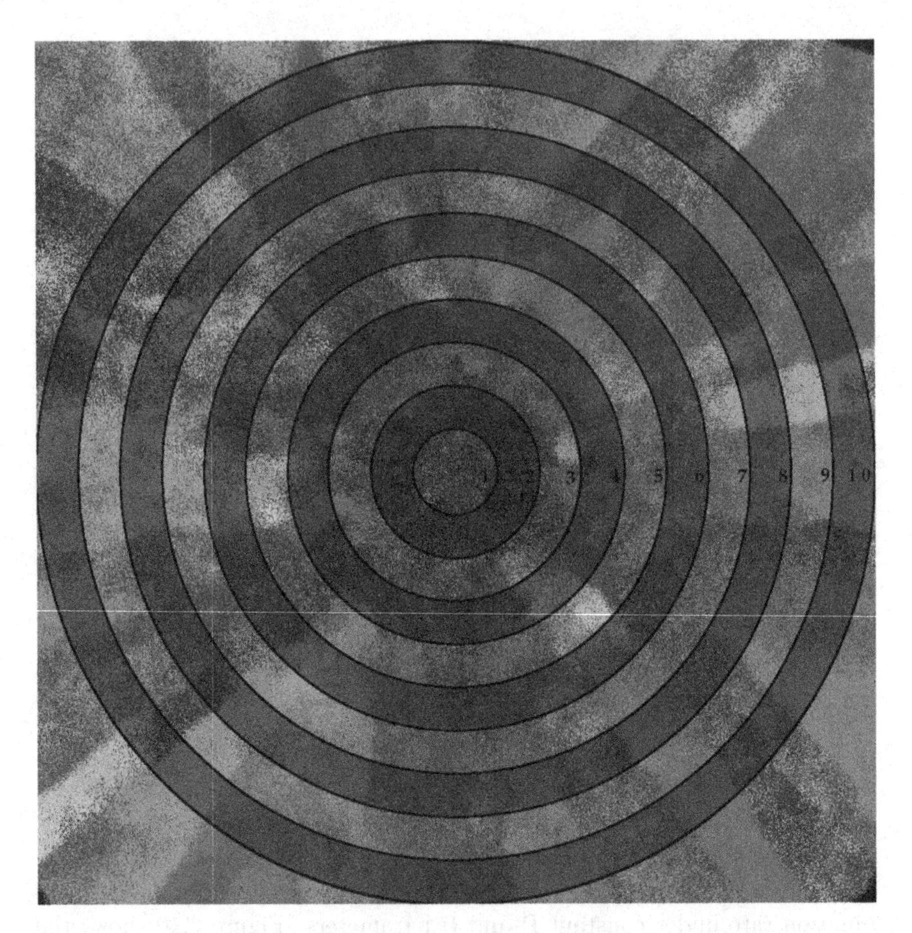

Fig. 3.11. The expanding population. Rings delimit "geographic regions" further analyzed for the defective gene distribution and the distribution of accepted crossover events (next two figures). Parameters of simulations: P=B=5, crossover rate C=0.1, L=64.

stay at the same place for whole their lives, only newborns can be put aside, in the new non occupied territories. But the newborns can be put also inside if there is a place. Those complicated relations should affect the evolutionary mechanisms operating inside the population and should be seen in the genetic pool of populations. We have analyzed the population evolving under parameters: lattice size 1024x1024, L=64, M=1, P=B=5, crossover rate 0.1. Figure 3.11 shows how the population looks in the early stages of evolution, still in an expansion. Figure 3.12 shows the "geographical distribution" of defective genes in the population. At the

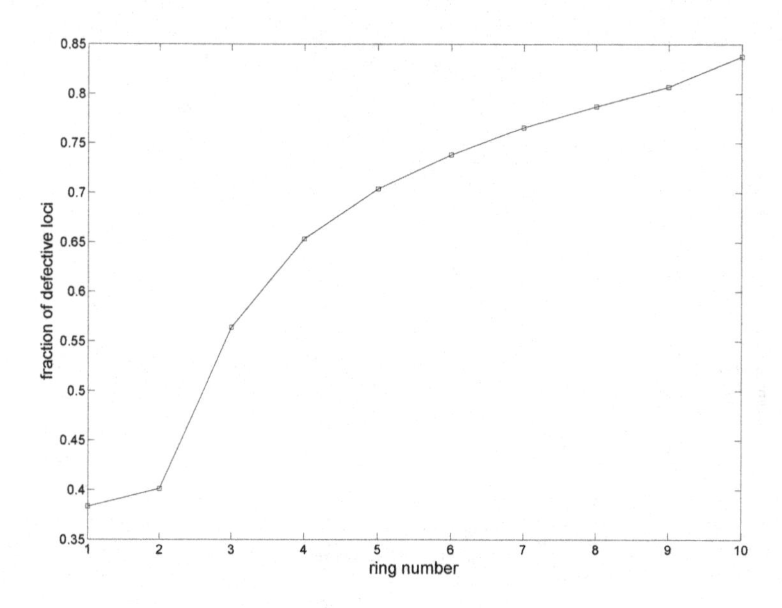

Fig. 3.12. Distribution of defective genes (number of heterozygous loci) in the expanding population. The fraction of defective genes is lower in the center of population, where for long time the whole lattice has been occupied and the inbreeding is lower. At the edges of expanding population inbreeding is higher and populations choose the complementing strategy.

edges of population inbreeding coefficient is higher and the complementing strategy can prevail, that is why higher fraction of defects is found in the outer parts of population. On the other hand, in the center of expanding populations individuals reproduce under lower inbreeding for longer time thus, the purifying strategy prevails in those parts of population and the lower fraction of defects is found in this region.

3.7.5. *Distribution of accepted crossover events*

In the population described in the above subsection the probability of recombination between haplotypes during gamete production was 0.1. The location of recombination was randomly chosen thus, the spots of recombination events should be evenly distributed along the chromosomes in the whole pool of gametes. Nevertheless, only some gametes form the surviving zygotes. That is why we have checked if in the surviving individuals the recombinations spots are evenly distributed, as expected. We have called

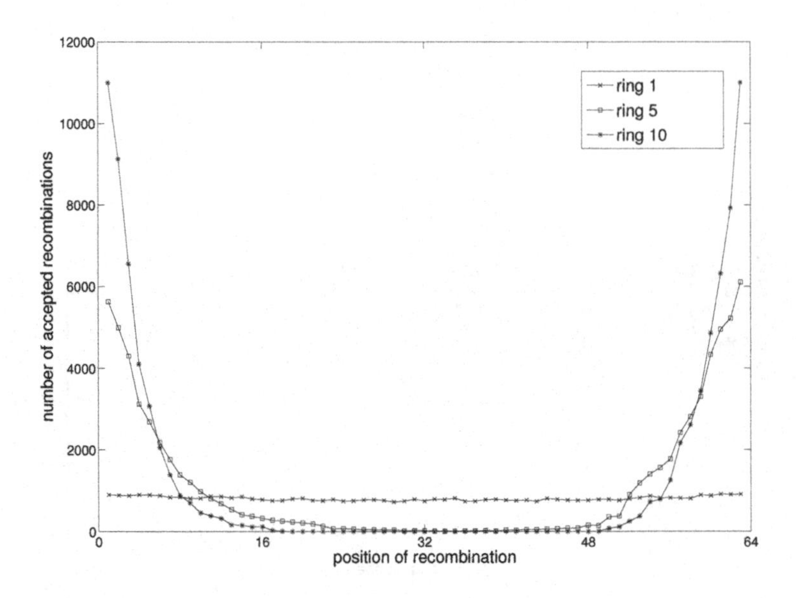

Fig. 3.13. Distribution of accepted recombination events along the bitstrings in different regions of population. The probability of recombination was independent on position thus, one could expect recombination events are distributed evenly. Accepted recombination means here that haplotype after recombination formed a surviving zygote. In the center of the population the recombination events are distributed evenly - as expected. At the regions more distant from the center the recombinations are accepted more frequently at the ends of bitstrings which correspond sub-telomeric regions of chromosomes.

these recombination events - "accepted recombinations", since the recombination product forms the surviving individual. We have checked the acceptance of recombinations in three parts: in the center, the fifth ring and the tenth ring. Plots in Fig. 3.13 show how these distributions depend on time and on the "geographical" location of individuals. During the first stages of evolution in the center of population the accepted recombinations were distributed evenly. Going to the outer parts of population the characteristic bias in distribution is observed - the frequency of acceptance the recombination is higher at the terminal regions of bitstrings (in genetic terminology - in sub-telomeric regions). In the most external parts of population the recombinations in the middle of bitstrings are not accepted - it means that all recombinations which happened in this part are lost in zygotic death. How it could be explained? In the genetic language two adjacent genes are linked and the recombination probability between them is low. If we think

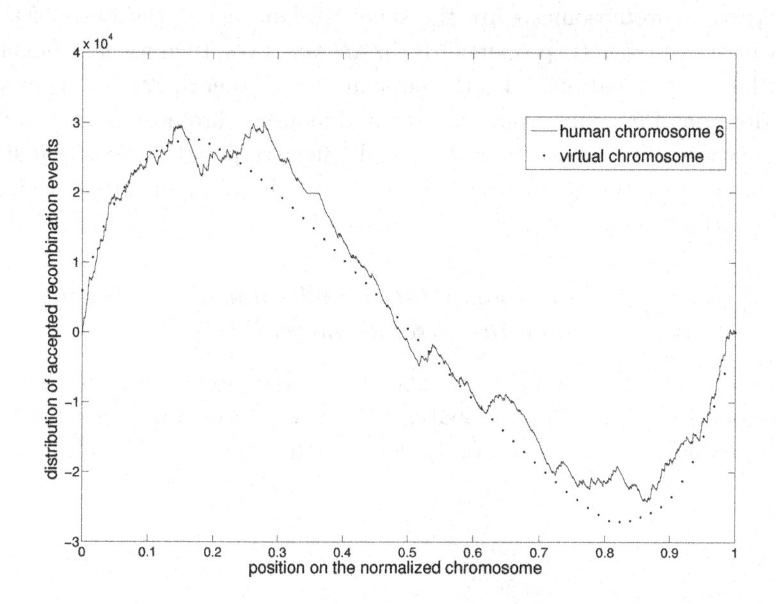

Fig. 3.14. Cumulative detrended plot of frequency of accepted recombination events along the human chromosome and along the virtual chromosome. In the upward directed regions of plots the recombinations are higher than average and in the downward directed regions - lower than average.

about the bitstrings in our model it is obvious that any recombination which happens between two bitstrings separates the first bits from the last ones located at the parental bitstrings. Closer to the center of bitstring - lower the probability of the separation of two bits by recombination somewhere between them. That is why the strategy of complementation is preserved in the centre of chromosomes longer than in the subtelomeric parts. It is a very interesting outcome of the model - the recombination events accepted by selection are not evenly distributed along the chromosomes of evolving populations. We have checked this property analyzing the genomic data bases of mammals. Figure 3.14 presents the detrended cumulative plots for virtual chromosome and for human chromosome. The plots were prepared by cumulating the data of frequency of recombinations in the region of chromosome diminished by the expected frequency of recombination in the region counted under assumption that recombinations are accepted evenly along the whole chromosomes or bitstrings. The distribution of accepted recombinations in the natural chromosomes resembles the distribution in the virtual chromosomes - are the same mechanisms at the basis of that very important genetic property? If the answer is yes, then we can conclude that human populations (also the populations of other mammals, since the distribution of recombinations in rat's and mouse's chromosomes are of the same type) have evolved in rather small effective populations which shift the strategy of the genome evolution toward the complementing at least some parts of their genomes.

3.7.6. *How selection shapes the distribution of recombination spots along the chromosomes?*

In the above section we have described one of the mechanisms which could be responsible for somehow global distribution of recombination spots. Recombination spots are not evenly distributed along the natural chromo-

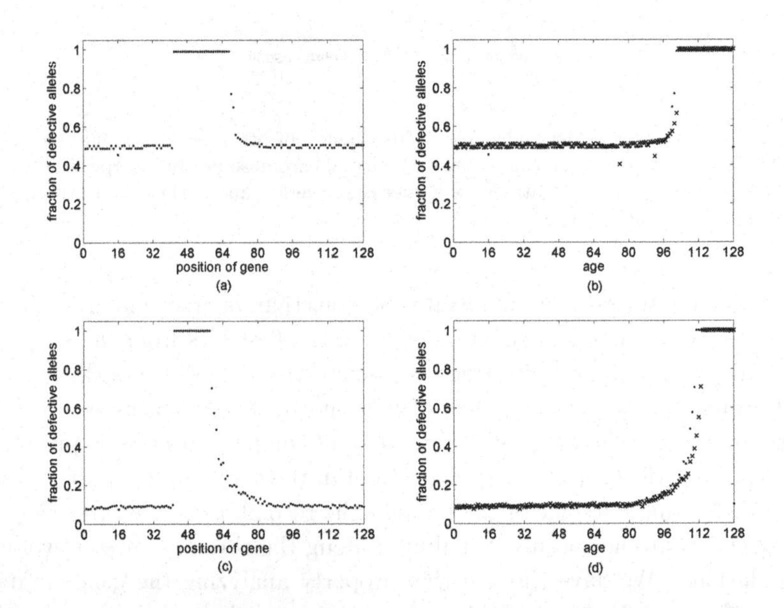

Fig. 3.15. Distribution of defective genes in the genetic pool of populations evolving with recombination frequency 1 per gamete production (low panel) and, without recombination (top panel). Left panel (a,b) - the regions of haplotypes from 42 -128 bits are inverted and switched on as indicated by numbers on x-axis; a,c - recombination C=0 , b,d - recombination C=1. Right panel (c,d), the data from the left panel compared with the standard structure of genomes (non-inverted). Now x-axis is scaled accordingly to the chronology of switching on; it is in agreement with the structure for standard model only.

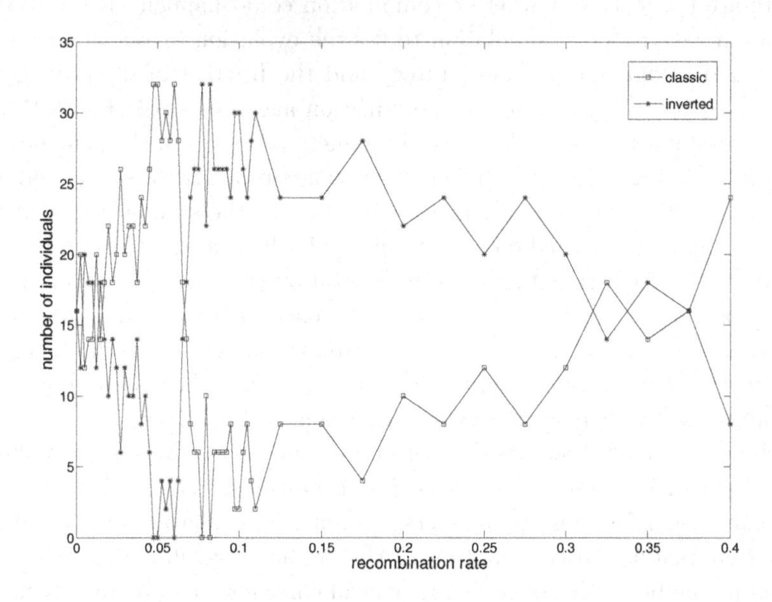

Fig. 3.16. Number of winning competitions between populations with standard genome structure and populations with "inverted" regions of genomes.

somes. There are some regions with very high recombination frequency (called hot-spots) and some regions with very low frequency of recombination (called recombination deserts) [12], [13], [14], [15]. Is it possible that selection would favor one configuration of genome over the other one only on the basis of distribution of frequency of recombination? To check such possibility we have performed some experiments with the Penna model. Reader has to refresh his knowledge of the Chap. 2. If the simulations of population evolution are performed under parameters N=10000, L=128, M=1, B=2, R=80 and crossover rate 1 per gamete production the distribution of defective genes is characteristic for the Penna model - low and constant frequency of defects in genes expressed before the minimum reproduction age R and fraction of defects growing with age after the minimum reproduction age. All genes in the genetic pool located at position above 110 are already defective (Fig. 3.15). Thus, the last 20 positions have no effect on surviving of the individuals - at the age when those genes should be switched on, the individuals are already dead. The space occupied by these genes, from the point of view of genetic information is empty. But there is still the "space", it corresponds to the region of chromosome where there are no genes but where recombination could happen. If the recombination rate is set in simulation to 0 - the evolution drives the genomes toward the complementation strategy and the fraction of defective genes expressed before the minimum reproduction age equals 0.5 (Fig. 3.15). If the recombination rate is high - 1 per gamete production, the genomes are under purifying selection. Imagine the genome of 128 bits long but the genes are switched on in different order than in the standard model; the first 41 bits are switched on at the same order but the next region (loci 42 - 128) is inverted, thus the locus 42 is positioned now at the other end of chromosome. Is there any significant difference in the evolutionary values of such structure of genomes when compared with the standard model? Comparing the results of simulations it shows that there is no difference between the two configurations of chromosomes (Fig. 3.15).

In simulations shown in Fig. 3.15 the evolution conditions (recombination frequencies) were far from critical conditions. The critical value of recombination for these parameters of simulations is around 0.076. What could happen at critical conditions? Under such recombination rate some genes at the both ends of coding regions of chromosomes are already under purifying selection. There is a "biological" experiment which could allow to estimate the evolutionary values of such populations. We can put the two populations into one environment and look for the winner. If populations have the same "evolutionary" value, the number of winning and loosing competitions for each population should be statistically the same. We have simulated 32 populations with a standard configuration of loci and 32 populations with the "inverted" second part of the genome and performed competitions of 32 pairs of such populations. The series of such 32 competitions were performed for populations evolved under different crossover frequency.

The results are shown in Fig. 3.16. For recombination rate close to 0 it was no difference between the two configurations, but already for recombination 0.02 the standard configurations won more often than the inverted one. For recombination rate in the range between 0.045−0.065 the standard populations won the most of competitions (118/128), inverted populations were losers. Situation changed dramatically between 0.06 and 0.075 - in the range of recombination frequency 0.0775 and 0.10 the most of populations with standard configurations lost the competitions (23/160).

Even more interesting is the distribution of defective genes in the genetic pool of winners and losers, shown in Fig. 3.17. At recombination rate of 0.06, the winning population had more defective genes expressed before

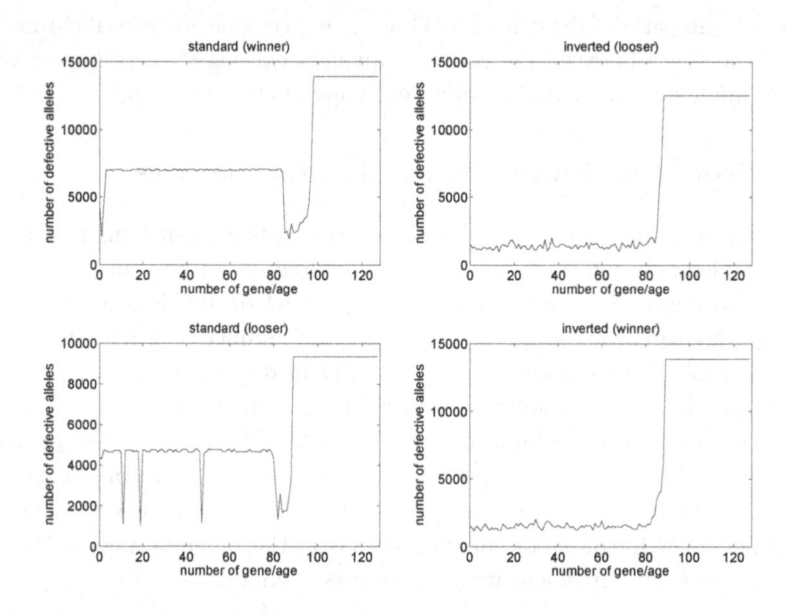

Fig. 3.17. The genetic structure of pairs of winning/loosing populations. The first pair (upper) evolved under recombination rate 0.06, the second pair under recombination rate 0.08.

the minimum reproduction age than the loosing one. But the winners complemented their haplotypes. Losers had less defective genes but they were already under the purifying selection, it was evidently more costly strategy than that of winners. Increasing the recombination rate up to 0.08 changes the situation substantially. Now the population under purifying selection is better than population using the complementary strategy. The last one is already trying to change the strategy and larger fractions of genomes are already under purifying strategy. Analyzing the results one has to notice that populations with inverted genome enter the purifying strategy earlier than the populations with standard genomes. Why? Because in their genomes, the defective genes which do not play any informational role (empty space) are located in the middle of the coding sequences. Any recombination inside this "empty space" reshuffles mutually the genes located at both ends of chromosomes and disturbs the possibility of complementation. On the other hand, in the non-inverted configuration, all recombinations which happened inside this genetically "empty space" had no effect on the complementation because they did not change the distribution of genes in the really coding parts of the genomes. One can expect that in natural chromosomes the space between the coding sequences can play a very important role in modeling the recombination landscape of chromosomes.

3.8. Phase transition and the length of chromosomes

In the subsection dealing with phase transition between purifying selection and complementation strategy we have presented data concerning one pair of chromosomes. We have used bitstrings of 64 or 100 bits long which corresponds to chromosomes coding the same number of genes. Critical frequency of recombination was of the order of 0.1 crossovers per pair of bitstrings which corresponds to 10 centi Morgans (cM) in genetic units (1 cM corresponds to probability of crossover 0.01). Considering the coding density as a number of genes per 1 cM we can estimate that in our simulations it was of the order 10 bits per 1 cM. In human genomes number of genes per 1 cM for different chromosomes is in the range between 2.5 and 13.6. It is a good agreement with our results of simulations. Generally the number of bits per 1 cM is growing with the length of chromosome but the slope depends on the size of population. If we assume that human chromosomes evolved under conditions described by our simulation then their coding density could also corresponds to our virtual chromosomes. We have put the data corresponding to human chromosomes (coding density in relation to the size of chromosomes in genes' number). All points representing the data are found for low effective populations. Does it mean that human population have evolved under high inbreeding?

3.9. Kinship and fecundity in the human population

Fecundity (called also the total fertility rate) is a specific, very accurate measure of fertility of a given population. It corresponds to the average number of offspring produced by one female. For human contemporary populations (European) it should be around 2.08 to fill up the gaps in our populations caused by natural and random deaths. Recently, the data concerning the fecundity of Icelander population were published [22]. The data covers the period from 1850 up to 1950. Authors discovered that there is a positive correlation between the fecundity and genetic relation between spouses. The highest fecundity was observed for the genetically high related spouses. The real evolutionary success measured by the number of grand children was found for spouses related at the third - fourth cousins' level.

Authors have claimed that they excluded any social and economical effects and only biological mechanisms should be responsible for their finding.

Let's try to analyze the phenomenon from the point of view of our model. The most important assumption: the higher fecundity could be related to lower zygotic death - lower probability that a zygote does not survive until its birth or, better, until its minimum reproduction age. We have observed the phase transition in the biological reproduction - there is a value of crossover rate where, for a given size of effective population, the fecundity is the lowest (the highest number of unsuccessful trials of producing the offspring per one survival, see section and Fig. 3.5). Thus, an effect we have observed is exactly reciprocal to the expected explanation of Islander fecundity - where a maximum of fecundity was observed. Nevertheless, the transition point was characteristic for a given effective size of population whose genomes were represented by single pairs of chromosomes. In the human genomes we have 23 pairs of chromosomes of different length and

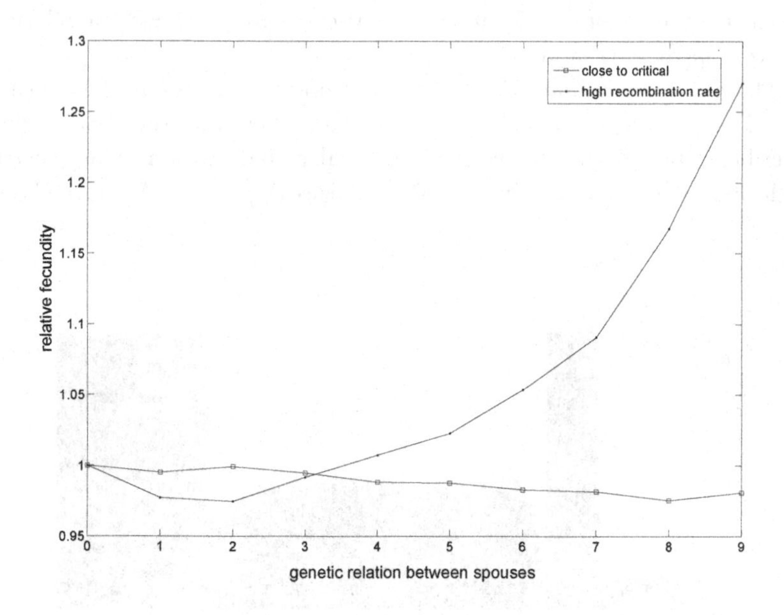

Fig. 3.18. The relationship between fecundity and genetic relations between parents (x-axis show the level of cousins' relations, 0 is denoted for siblings). Growing plot is for recombination rate much higher than critical transition point, decreasing plot show the fecundity for recombination rate below the critical value for longer chromosome and above the critical value for shorter chromosome.

coding density per recombination units. Especially the last value is an important parameter establishing the transition point.

Let's assume the model where individuals are represented by genomes composed of two pairs of chromosomes with different coding density per recombination unit. The critical recombination value for each pair should be different, thus for a given and constant condition of recombination the critical inbreeding (the effective population size) should be different for these chromosomes. Though, each single chromosome tends to keep the population out of its transition point and it is possible that the optimum effective size of evolving population should be somewhere in-between the critical values. If that reasoning is correct, we have to find such optimum in the model.

If the individual genomes in panmictic population (N=1000) are composed of two bitstrings L1=64 and L2=256, at the recombination frequency 0.3 the shorter chromosome is under purifying selection while the longer one under complementation strategy. In such conditions the fecundity of populations increases with the inbreeding. If the recombination is set to 1 per chromosome pair, the fecundity is anti-correlated with genetic relations between partners (Fig. 3.18).

The other preliminary data have been obtained in simulations on lattice. The parameters of the model have been chosen on the basis of data describing the relation between the critical recombination values, length of chromosomes and effective population sizes (Fig. 3.5). We have chosen

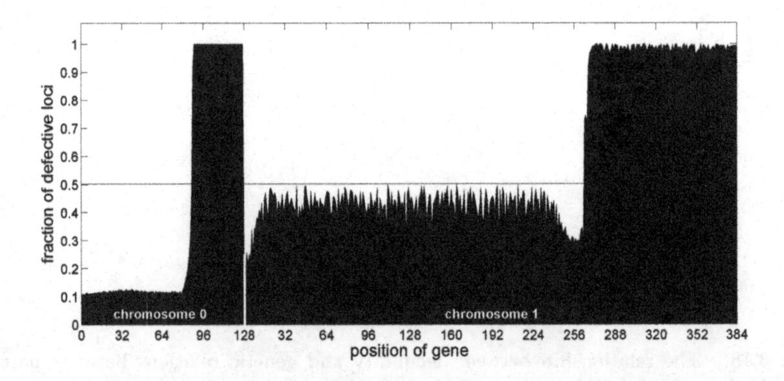

Fig. 3.19. Genetic pool structure for population evolving on lattice with one chromosome 128 bits long and the second one for 384 bits long. Other parameters: recombination rate 0.24 for each pair, P=B=4.

chromosome pairs L1=128 and L2=384, the square lattice size 512x512 and recombination rate 0.24, P=B=4. Under such conditions the shorter chromosome stays under purifying strategy while the longer one under complementary strategy (Fig. 3.19). Further studies are connected with looking for conditions where non-monotonous relation for real evolutionary costs could be obtained. We expect that in shrinking population (higher inbreeding) selection would try to push the shorter chromosome under complementing strategy (bad situation - higher zygotic mortality). Increasing the effective population (lower inbreeding) would try to push the longer chromosome toward the purifying selection - also bad situation, again higher zygotic death frequency. What is the solution? To stay with the optimum of inbreeding - population can grow but the traditional reproduction habits should be kept or restored - they supposed to be the best.

Acknowledgements

The work was done in the frame of European programs: COST Action MP0801, FP6 NEST - GIACS and UNESCO Chair of Interdisciplinary Studies, University of Wrocław. Calculations have been carried out in Wrocław Centre for Networking and Supercomputing (http://www.wcss.wroc.pl), grant # 102.

References

[1] E. Mayr, *Systematics and the Origin of Species.* (Columbia Press, New York, 1942).

[2] C. D. Jiggins, Sympatric speciation: why the controversy?, *Curr. Biol.* **16**, R333—R334, (2006).

[3] M. Barluenga, K. N. Stolting, W. Salzbulger, M. Muschick and A. Meyer, Sympatric speciation in Nicaraguan crater lake cichlid fish, *Nature* **439**, 719—723, (2006).

[4] P. M. E. Bunje, M. Barluenga and A. Meyer, Sampling genetic diversity in the 16 sympatrically and allopatrically speciating Midas cichlid species complex over a 16 year time series, *BMC Evol. Biol.* **7**, 25—39, (2007).

[5] M. Doebeli and U. Dieckmann, Speciation along environmental gradients, *Nature* **421**, 259—-264, (2003).

[6] D. Stauffer, S. Moss De Oliveira, P. M. C. De Oliveira and J. S. Sa Martins, *Biology, Sociology, Geology by Computational Physicists.* (Amsterdam: Elsevier, 2006).

[7] P. M. C. De Olivera, S. Moss De Oliveira, D. Stauffer, S. Cebrat and A. Pękalski, preprint: Does sex induce a phase transition?, arXiv:0710.1357

[8] T. J. P. Penna, A bit-string model for biological aging, *J. Stat. Phys.* **78**, 1629—1633, (1995).

[9] J. W. Drake, B. Charlesworth, D. Charlesworth and J. F. Crow, Rates of spontaneous mutation. *Genet.* **148**, 1667—1686, (1998).

[10] M. YA. Azbel, Phenomenological theory of mortality evolution, *Proc. Natl. Acad. Sci. U.S.A.* **96**, 3303—3307, (1999).

[11] S. Cebrat and D. Stauffer, Gamete recognition and complementary haplotypes in sexual Penna ageing model, *Int. J. Mod. Phys. C.* **19**(2), 259—265, (2008).

[12] A. Yu et al., Comparison of human genetic and sequence-based physical maps, *Nature* **409**, 951—-953, (2001).

[13] M. J. Daly, J. D. Rioux, S. F. Schaffner, T. J. Hudson and E. S. Lander, High-resolution haplotype structure in the human genome, *Nature Genet.* **29**, 229—232, (2001).

[14] A. J. Jeffreys, J. Kauppi and R. Neumann, Intensely punctate meiotic recombination in the class II region of the major histocompatibility complex, *Nature Genet.* **29**, 217—222, (2001).

[15] N. Arnheim, P. Calabrese and M. Nordborg, Hot and cold spots of recombination in the human genome: the reason we should find them and how this can be achieved, *Am. J. Hum. Genet.* **7**, 5—16, (2003).

[16] K. Bońkowska, M. Kula, S. Cebrat and D. Stauffer, Inbreeding and outbreeding depressions in the Penna model as a result of crossover frequency, *Int. J. Mod. Phys. C.* **18**, 1329—1338, (2007).

[17] M. Zawierta, P. Biecek, W. Waga and S. Cebrat, The role of intragenomic recombination rate in the evolution of population's genetic pool, *Theory in Biosciences*, doi:10.1016/j.thbio.2007.02.002, (2007).

[18] W. Waga, D. Mackiewicz, M. Zawierta and S. Cebrat, Sympatric speciation as intrinsic property of expanding populations, *Theory in Biosciences* **126**, 53—59, doi:10.1007/s12064-007-0010-z, (2007).

[19] B. Spyropoulos, P. B. Moens, J. Davidson and J. A. Lowden, Herozygote advantage in Tay-Sachs carriers?, *American Journal of Human Genetics* **3**, 375—380, PMID 7246543, (1981).

[20] J. Hardy and H. Harpending, Natural History of Ashkenazi Intelligence. PDF Retrieved on January 29 (2006).

[21] D. Kwiatkowski, Genetic susceptibility to malaria getting complex, *Current Opinion in Genetics & Development* **10**, 320—324, (2000).

[22] A. Helgason et al., An association between the kinship and fertility of human couples, *Science* **319**, 813, (2008).

[23] `http://www.smorfland.uni.wroc.pl/sympatry/`

Chapter 4

Models of population dynamics and their applications in genetics

Ryszard Rudnicki

Institute of Mathematics, Polish Academy of Sciences,
Bankowa 14, 40-007 Katowice, Poland
and Institute of Mathematics, Silesian University,
Bankowa 14, 40-007 Katowice, Poland,
rudnicki@us.edu.pl

The aim of these lectures is to give a survey of mathematical models and methods of population dynamics. The main subject will be structured models which describe the distribution of population with respect to some parameters such as age, size, maturity or proliferative state of cells. Some models are general, *e.g.* age-structured model can be applied to human population as well as to a cell population but others will concern specific populations *e.g.* the population of erythrocytes. We concentrate on models connected with genetics: the Penna model and models describing telomere shortening, distribution of genes in a genome, stochastic gene expression and the cell-cycle.

Contents

4.1. Introduction

Population dynamics is the study of changes in the number and composition of individuals in a population, and the factors that influence those changes. Although the first population models appeared in demography and later in ecology and epidemiology, they have become increasingly important in almost all branches of biology. Methods of population dynamics are applied in ecology, epidemiology and infectious diseases, genetics, physiology, immunology and cancer growth. The rapidly developing techniques of molecular biology and genetics produce large quantities of data, that demand mathematical analysis and modeling. Using mathematical models one can analyze populations at various levels, including cells, genes, and biomolecules. Nowadays mathematical modeling of population dynamics is a central topic in theoretical biology and some biologists find that mathematical models are absolutely essential for research in modern biology. Mathematics provide a broad spectrum of methods to study population dynamics. The models use all types of differential equations, probability theory, dynamical systems, discrete mathematics and also very complicated systems which include age, stage or size structures.

Modern genetics use a wide spectrum of mathematical models which describe the gene distribution in evolving populations, changes in a single genome or biochemical processes regulated by genes. Though genetics concerns small biological objects, the methods and models of the traditional population dynamics can be successfully applied to study genetic problems. On the other hand some genetics models are completely original, *e.g.* such as the Wright-Fisher and Moran models, and they initiate intensive development of the population dynamics.

Since a population is usually heterogeneous it is important to divide the population into homogeneous groups according to some significant parameters such as age, size, maturity or proliferative state of cells and study interactions between such groups. Models of this type are called *structured* and they describe the time evolution of the distribution of the population according to the fixed parameters. Structured model appeared for the first time in demography, where they described the age structure of human populations, but now they are used in all fields of biology. There are hundreds of articles and several books and collections of articles devoted to structured population models (see *e.g.* [1–8]).

The aim of these lectures is to give a survey of mathematical models and methods of population dynamics. Some models are general, *e.g.* the age-structured model can be applied to human population as well as to a cell population but others will concern specific populations *e.g.* the population of erythrocytes. The main subject will be structured populations but our aim is not to give a comprehensive review of this subject matter or even a selected part of it. We dedicate this article to students who can feel completely lost in enormous numbers of models and results concerning this subject. We want to show that most of models can be easily derived if we know only a few basic models and some general roles of investigation of the problem. Since a good model should give clearly formulated conclusions we show that our models really have this property. Most of them have asynchronous exponential growth — the behavior having a simple biological interpretation. But since reality is not so simple we also explain when we can expect other behavior. In order to achieve such aims we arrange the material according to mathematical structure. We consider continuous time models and discrete time models which usually describe the relations between consecutive generations. We also distinguish models according to the type of structure (discrete or continuous). We divide the lecture into four parts considering one type of models in one part. The theory of structured models is preceded by a section devoted to the history of mathematical modeling. In this section we present "classic models", but these models are still living and can be used as starting points in the investigations of complex phenomena. If we know the properties of a simplified model we can predict what can happen in the complex version. We will finish with a short conclusion section which provides possible directions of future investigations. Models connected with genetics play the crucial role in our presentation, although we do not collected them in one or two sections. We consider here the Penna model and models describing telomere shortening, distribution of genes in a genome, stochastic gene expression and the cell-cycle.

4.2. History of mathematical modeling in biology

In many studies the history of mathematical modeling in biology usually begins in 1202. In that year Fibonacci in the book *Liber Abaci* introduced the sequence, further named after him, to describe the growth of a rabbit population. But the real history is a bit different. First, this sequence was known and used by Indian mathematicians in 500 B.C. Second, more interesting and surely more important from the view of modern science was Ulpian's table of life expectancies [9, 10] which dates from about 220 A.D. In some sense Ulpian's table can be treated as the first age-structured model because it is possible to deduce from this table the age distribution of the Roman population.

Until the XX century demography was the only part of life sciences which used mathematics. We mention here a few names connected with these investigations. E. Halley, the most known as the astronomer, published in 1693 two articles [11, 12] on life annuities based on the mortality tables for the city of Wrocław (at that time Breslau). This work was highly influential on the development of life insurance. In 1798 T. Malthus claimed in his book [13] that the human population will grow exponentially, *i.e.* according to the equation $N'(t) = \lambda N(t)$. Here and further $N'(t)$ denotes the derivative dN/dt. It is interesting that the old Malthusian model works properly not only in a biological laboratory. In some periods of time the world population really grew exponentially. For example, according to data taken from [14] in years 1950-1985 this growth was almost exponential and the doubling time $T = \ln 2/\lambda$ was about 36 years, but now the growth rate is a half of that one.

B. Gompertz in [15] improved in 1825 the Malthus model and proposed that the number of individuals at time t satisfies the following differential equation

$$N'(t) = \lambda N(t) \log(N(t)/K). \tag{4.1}$$

P. F. Verhulst [16] proposed in 1838 a similar model

$$N'(t) = \lambda(1 - N(t)/K)N(t). \tag{4.2}$$

Both models are based on the assumptions that the available resources are limited, per capita birth and mortality rates depends on the population size and there is an optimal population size K. They are simple modifications of the Malthus model. The constant *per capita growth rate* λ in the Malthus model was replaced with the growth rates which depends on the population size: $f(N) = \lambda \log(N/K)$ in the Gompertz one and $f(N) = \lambda(1 - N/K)$ in the Verhulst one. If the initial size of the population is less than K, then the population grows and converges to its maximal size K. But in some situations such as an invasion on a new territory or sudden changes in the environment the initial size can be greater than K and then the population decreases and also converges to its equilibrium K. The Verhulst model is very popular and (4.2) is called the *logistic equation.*

W. C. Allee [17] observed that the reproduction and survival of individuals decrease for smaller populations and it can lead to the extinction of the population. This phenomenon is called the *Allee effect* and the reason is that small density of the population has a negative influence on the reproduction and survival of an individual. Moreover, the small genetic diversity in such a population also gives a negative effect on its survival and adaptability. The Allee effect can be added to the Verhulst model by modifying its growth rate. Since the Allee effect decreases with the growth of the population size, the simplest way is to subtract from the growth rate the term $A/(1+BN)$ and, as a result, the size of the population can be described by the equation

$$N'(t) = \lambda\left(1 - \frac{N(t)}{K} - \frac{A}{1+BN(t)}\right)N(t). \tag{4.3}$$

The real progress in the development of mathematical models in biology began in the 1920's. In that time V. Volterra [18, 19] studied the question: why did a complete closure of fisheries during the First World War cause an increase in predatory fish and a decrease in prey fish in the Adratic Sea? He proposed a model which describes the relation between the number of prey $N_1(t)$ and predators $N_2(t)$. This model consists of the following system of differential equations

$$\begin{cases} N_1'(t) = (\varepsilon_1 - \gamma_1 N_2(t))N_1(t), \\ N_2'(t) = (-\varepsilon_2 + \gamma_2 N_1(t))N_2(t). \end{cases} \tag{4.4}$$

A similar model was proposed independently by A. J. Lotka [20] and it is today known as the Lotka–Volterra prey-predator model. The system has one positive equilibrium (K_1, K_2), $K_1 = \dfrac{\varepsilon_2}{\gamma_2}$, $K_2 = \dfrac{\varepsilon_1}{\gamma_1}$ and other solutions are periodic functions with some period T dependent on the initial data. But even more interesting property of the model is that the mean values of the number of prey and predators are constants:

$$\frac{1}{T}\int_0^T N_i(t)\,dt = K_i. \tag{4.5}$$

Observe, that from (4.5) it follows that if both prey and predators are fished then $\varepsilon_1' < \varepsilon_1$ i $-\varepsilon_2' < -\varepsilon_2$ and

$$K_1' = \frac{\varepsilon_2'}{\gamma_2} > \frac{\varepsilon_2}{\gamma_2} = K_1,$$

$$K_2' = \frac{\varepsilon_1'}{\gamma_1} < \frac{\varepsilon_1}{\gamma_1} = K_2,$$

where $'$ corresponds to the model including fishing. It means that if both populations are fished then the mean size of the population of prey grows and the population of predators decreases. The opposite process took place during the First World War. The Lotka-Volterra model was generalized in many ways. One of them was the Kolmogorov model [21] given by the system of equations

$$\begin{cases} N_1'(t) = \varepsilon_1(N_1(t))N_1(t) - \nu(N_1(t))N_2(t), \\ N_2'(t) = \varepsilon_2(N_1(t))N_2(t). \end{cases}, \tag{4.6}$$

where ε_1 is a decreasing function, ε_2 is an increasing function and ν is a positive function. If we choose properly functions ε_1, ε_2, and ν then the system has a limit cycle. We recall that the *limit cycle* is a periodic solution such that the graphs of other solutions in phase space spiral into its graph as $t \to \infty$. This property is very interesting from both mathematical and biological point of view. It denotes that there exists one periodic solution and the system is not sensitive on small perturbations – after sufficiently large time the size of the coexisting prey and predator populations will return to the periodic state. The book [22] contains an interesting collection of papers from those time concerning ecology. Prey-predator models are still intensively investigated (see [23]).

The second big impulse in the development of mathematical models in biology were A. G. McKendrick works. He created with W. O. Kermack the mathematical theory of epidemics with the famous SIR model [24]. In this model the population is split into three groups whose numbers are denoted by S - susceptible, I - infected, and R - resistant (or removed). The model based on the following assumptions:

1) a susceptible individual who becomes infected goes immediately to the second group I and it can transmit the disease,
2) an infected individual can become resistant by recovery or quarantine and it become permanently immune; we include in the resistant group the dead individuals,
3) we neglect all demographic processes, the total number of individuals $N = S + I + R$ is constant,
4) the disease is transmitted directly, *i.e.* there is no intermediate host, *e.g.* mosquito for malaria,
5) the population is homogeneous — with the same probability each infected individual can infect a susceptible one.

The model is the following

$$\begin{cases} S'(t) = -\alpha S(t)I(t), \\ I'(t) = \alpha S(t)I(t) - \beta I(t), \\ R'(t) = \beta I(t), \end{cases} \tag{4.7}$$

where α is the rate at which a susceptible individual become infected by an infected one and β is the recovery (removal) rate. From the model it follows that there is no epidemic outbreak if and only if $S(0) \leq \frac{\beta}{\alpha}$. It is a practical information because it tells us how many people should be vaccinated to prevent the epidemic outbreak.

McKendrick [25] also introduced in 1926 a model which consists of a partial differential equation and a boundary integral condition and describes the age distribution of a population. This model will be discussed in details in Sec. 4.6. It should be noted that a similar model was introduced earlier by Sharpe and Lotka [26]. The Sharpe-Lotka-McKendrick model is one of the earliest models of the structured population dynamics. From now we restrict ourselves to structured models. Readers interested in other applications of mathematics in biology, especially ordinary differential equations are referred to the books by Murray [27], Thieme [28], Hofbauer and Sigmund [29], Brauer and Castillo-Chavez [30], and Farkas [31].

In these lectures we present only selected models from genetics and we restrict ourselves only to deterministic ones. The systematic study of mathematical theories of population genetics can be found in the books by Crow and Kimura [32], Ewens [33], Fisher [34], Hartl and Clark [35], Kimura [36], Moran [37], Nagylaki [38]. An overview of stochastic models in genetics can also be found in Blythe and McKane [39].

4.3. Discrete models

In this section we consider the simplest structured models where both time and the structure are discrete. Such models are called the *Leslie models* [40]. The population is divided into n subpopulations, but sometimes it is convenient to consider infinite number of subpopulations. We consider disjoint generations of individuals. The time is discrete and is identified with generations, so the k-th generation lives at time k. We should underline that the word "generation" used here has a general meaning and only in particular cases it is a real generation. We assume that an individual from the j-th subpopulation "produces" with probability $p_{i,j}$ a new individual in the i-th subpopulation. Let x_i^k be the number of individuals in the subpopulation i in the k-th generation. Let P be the matrix with entries p_{ij}, and let $\mathbf{x}^k = [x_1^k, \ldots, x_n^k]$ be the vector-column of the age distribution in the population at time k. Then

$$\mathbf{x}^{k+1} = P\mathbf{x}^k. \tag{4.8}$$

We should underline that $[x_1^k, \ldots, x_n^k]$ do not need to be a probability vector and the sum $N^k = x_1^k + \cdots + x_n^k$ can depend on k. Here N^k is the population size in the k-th generation.

Example 4.1 (discrete age-structured model). *Here a generation is the set of individuals at a given time k, e.g. a year. Let the i-th subpopulation consists of individuals with age $i \le a < i+1$. Then $p_{j+1,j} = 1 - \mu_j$, where μ_j is the death rate at age j and $p_{0,j} = b_j$, where b_j is the birth rate at age j. In other cases $p_{i,j} = 0$.*

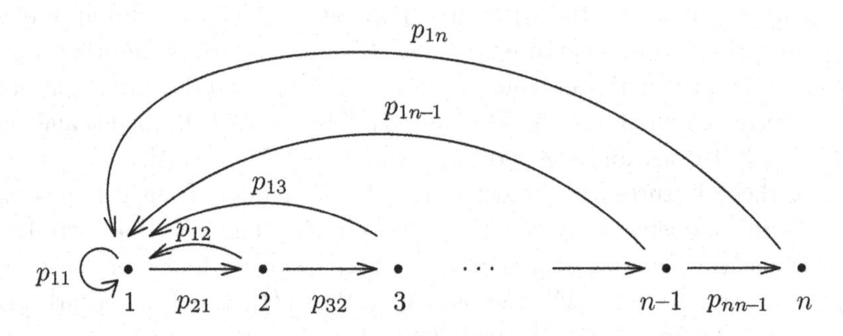

Fig. 4.1. The diagram of connections in the age-structured model. Here $p_{j+1,j} = 1-\mu_j$ and $p_{1,j} = b_j$, where μ_j is the death rate and b_j is the birth rate at age j.

Example 4.2 (space structured model). *As before a generation is the set of all individuals at a given time k and a subpopulation consists of all individuals living in a given region. It is sensible to assume that in this model $p_{i,j} > 0$ for all i, j.*

Example 4.3 (telomere shortening). *The ends of chromosomes, called telomeres, shorten when a cell divides. When they reach a critical length no further divisions occur. The simplified model of telomere loss is the following. Assume that the length of a telomere is an integer from the interval 0 to n. The i-th subpopulation consists of cells with a given telomere of the length i. We assume that a cell from the i-th subpopulation can die with probability μ_i or divide and its daughter cell can have the telomere with the length i with probability a_i or $i-1$ with probability $1-a_i$. The cells which belong to 0-th subpopulation cannot divide and they finally die. Let $r_i = 2a_i(1-\mu_i)$ and $d_i = 2(1-a_i)(1-\mu_i)$. The mean number of descendants of a cell from the i-th subpopulation in the same subpopulation is r_i and in* *the $(i-1)$-th subpopulation is d_i. In this case the matrix P is the following*

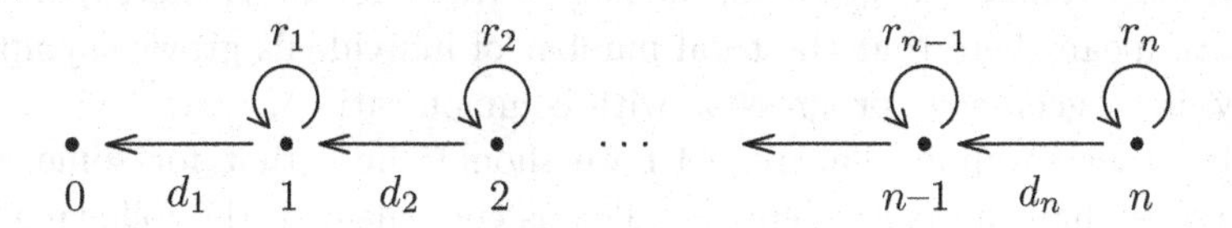

Fig. 4.2. The diagram of connections in the telomere shortening. In this case $p_{i,i} = r_i$ and $p_{i-1,i} = d_i$.

$$P = \begin{bmatrix} 0 & d_1 & 0 & \cdots & 0 \\ 0 & r_1 & d_2 & \ddots & 0 \\ 0 & 0 & r_2 & \ddots & \ddots \\ \vdots & \ddots & \ddots & \ddots & d_n \\ 0 & 0 & \cdots & 0 & r_n \end{bmatrix}.$$

The behavior of the vector $\mathbf{x}^k$ when time k goes to infinity is described by the following theorem.

Theorem 4.1 (Perron). *Assume that the matrix P has the non-negative entries and for some positive integer r the matrix P^r has all positive entries. Then there exist a constant $\lambda > 0$ and sequences $\mathbf{x}^* = (x_1^*, \ldots, x_n^*)$ and $\mathbf{y}^* = (y_1^*, \ldots, y_n^*)$ with positive terms such that for each $\mathbf{x} \in \mathbb{R}^n$ we have*

$$\lim_{k\to\infty} \lambda^{-k} P^k \mathbf{x} = \mathbf{x}^* \langle \mathbf{y}^*, \mathbf{x} \rangle, \tag{4.9}$$

where $\langle \cdot, \cdot \rangle$ is the scalar product in $\mathbb{R}^n$.

For the proof see [41] Proposition I.2.3.

If condition (4.9) holds then we say that the population has *asynchronous exponential growth.* Let us explain the meaning of this notion. The vector $\mathbf{x}^*$ can be chosen in such a way that $x_1^* + \cdots + x_n^* = 1$. The vector $\mathbf{p}^k = (p_1^k, \ldots, p_n^k)$, where

$$p_i^k = \frac{(P^k x)_i}{\sum_{j=1}^n (P^k x)_j} \tag{4.10}$$

describes the probability distribution of the population in the k-th generation. From (4.9) it follows immediately that

$$\lim_{k\to\infty} \mathbf{p}^k = \mathbf{x}^*. \tag{4.11}$$

It means that independently on the initial distribution $\mathbf{p}^0$ the long-time distribution is almost $\mathbf{x}^*$ (asynchronous growth). Moreover, the exponential growth means here that the total number of individuals grows asymptotically like a geometric progression with common ratio λ.

In order to apply Theorem 4.1 we should check that for some r the matrix P^r has all positive entries. This is equivalent to the following condition:

(c-d) for each i and j there exists a sequence $i_0, i_1, \dots, i_r$ such that $i_0 = j$, $i_r = i$ and

$$p_{i_r i_{r-1}} \cdots p_{i_2 i_1} p_{i_1 i_0} > 0. \tag{4.12}$$

Condition (4.12) denotes that if we start from the j-th subpopulation we can reach the i-th subpopulation in r steps going on the oriented graph from the diagram of the connections between subpopulations.

Let us return to our examples. In the space structure model (see Example 4.2) all entries of P are positive and consequently the population has the asynchronous exponential growth. More interesting situation appears in the age-structured model (see Example 4.1). If we assume that $b_j > 0$ for all $j = 1, 2, \dots, n$ then using the diagram of the connections one can check that the matrix P^n has all positive entries and we have the asynchronous exponential growth. But usually younger individuals cannot have offspring so we cannot assume that all b_j are positive. It turns out that it is sufficient to assume that $b_n > 0$ and $b_{n-1} > 0$. Then we can check that the matrix P^{n^2} has all positive entries. In some species of insects the life cycle can be constant and long. Thus we do not have the asynchronous exponential growth which can help the species to survive. For example, *the Magicicada* goes through a 17– or occasionally 13–year life cycle [42]. The predators are not able to adjust to this type of life cycle of cicadas. In the telomere shortening model (Example 4.3) we do not have the asynchronous exponential growth but the population has the *asynchronous polynomial exponential growth.* It means that for each $\mathbf{x}$ there exist constants $c_i(\mathbf{x})$ such that

$$x_i^k \approx k^{n-i} \lambda^k c_i(\mathbf{x}) \tag{4.13}$$

for large k.

We can also consider nonlinear Leslie models. Such models are defined by the formula $\mathbf{x}^{k+1} = \mathbf{f}(\mathbf{x}^k)$, where the function $\mathbf{f}$ is defined on a subset D of $\mathbb{R}^n_+$ which is invariant with respect to $\mathbf{f}$, *i.e.* $\mathbf{f}(D) \subset D$. The theory of such models is difficult and, in practice, each model is treated independently. We can study here classical questions such as stability, the existence of a limit cycle and also newer problems such as invariant measures and chaos. We give here only some examples which show how general assumptions concerning the population can be included in previous models.

Example 4.4 (age-structured model with limited resources). *Now we include in the age-structured model (see Example 4.1) the assumption on the limited resources which appears in the Verhulst model. Let $x(a,t)$ be the number of individuals with age a at time t, where both time and age are positive integers. Let $N(t) = \sum_{a=1}^{\infty} x(a,t)$ be the total number of individuals at time t. We can assume that the mortality rate μ depends on the age of an individual and the size of the whole population at given time, i.e. $\mu = \mu(a, N(t))$, $0 \le \mu \le 1$. We also assume that $b = b(a, N(t))$ is the mean number of children at time t of an individual with age a. For example, if we forget about the Allee effect we can take $\mu = N(t)/N_{\max}$ and $b = (1 - N(t)/N_{\max})p(a)$, where $N_{\max}$ is the maximum size of the population and $p(a)$ is the mean number of children in the best conditions. The general model is described by the following equations:*

$$\begin{aligned} N(t) &= \sum_{a=1}^{\infty} x(t,a), \\ x(t+1, a+1) &= (1 - \mu(a, N(t)))x(t,a), \quad \text{for } a \ge 1, \\ x(t,1) &= \sum_{a=1}^{\infty} b(a, N(t))x(t,a). \end{aligned} \tag{4.14}$$

If we assume that the maximum age is $a_{\max}$, then the infinity in the system (4.14) should be replaced by $a_{\max}$ and we should also assume that $\mu(a_{\max}, N) = 1$.

Example 4.5 (Penna model). *T. J. P. Penna [43] introduced a bitstring model for biological ageing which has been successfully applied to study various problems of genetics and demography. We present an analytical version of this model restricted to asexual reproduction [44]. In the Penna model it is assumed that each individual has its own maximum life span m. Let $x(t,a,m)$ be the number of individuals at time t with age a and with maxi-*

mum life span m. Then the model is described by the system of equations:

$$\begin{aligned} N(t) &= \sum_{m=1}^{\infty}\sum_{a=1}^{m} x(t,a,m), \\ x(t+1,a+1,m) &= (1-\mu(a,m,N(t)))x(t,a,m), \quad \textit{for } a<m, \\ x(t+1,a+1,m) &= 0, \quad \textit{for } a \ge m, \\ x(t,1,m) &= \sum_{m=1}^{\infty}\sum_{a=1}^{m'} b(a,m,m',N(t))x(t,a,m'). \end{aligned} \tag{4.15}$$

As in the previous model μ is the death rate, and $b(a,m,m',N)$ is the mean number of children with the maximum life span m of an individual with age a and maximum life span m'. In [44] it is assumed that $\mu = N(t)/N_{\max}$ and $b = (1 - N(t)/N_{\max})p(m,m')$, where $N_{\max}$ is the maximum size of the population and $p(m,m')$ is the birth matrix $p(m,m')$, i.e. $p(m,m')$ is the probability that a parent with maximum life span m' gives birth to a child with maximum life span m.

4.4. Time continuous discrete structure models

Now we consider models in which time has the ordinary physical meaning and the population is divided into a finite or infinite number of subpopulations. Models of this type appear in many applications. They can describe, for example, the migration of animals, birth and death processes, distribution of genes (we give examples in this section) and spreading diseases in heterogeneous populations (see [28] Chapter 24). The general scheme for such models is the following. In time interval from t to $t+\Delta t$ an individual from the subpopulation j can:

(a) "move" with probability $p_{ij}\Delta t + o(\Delta t)$ to the subpopulation i,

(b) "produce" with probability $b_{ij}\Delta t + o(\Delta t)$ a new individual in the subpopulation i,

(c) die with probability $d_j\Delta t + o(\Delta t)$.

Let $x_i(t)$ be the number of individuals in the subpopulation i at time t. Then

$$x_i'(t) = \sum_{j=1}^{n} q_{ij}x_j(t), \quad \text{for} \quad i = 1,\dots,n, \tag{4.16}$$

where $q_{ij} = b_{ij} + p_{ij}$ for $i \neq j$ and

$$q_{ii} = b_{ii} - d_i - \sum_{\substack{j=1 \\ j \neq i}}^{n} p_{ji}.$$

For $i \neq j$ we have $q_{ij} \geq 0$. If we consider only finite number of subpopulations then we can formulate a continuous version of the Perron theorem. We need the following condition:
(c-c) for $i \neq j$ there exists a sequence $(i_1, i_2, \ldots, i_m)$ such that $i_1 = i$, $i_m = j$ and $q_{i_{r+1}, i_r} > 0$ for $r = 0, 1, \ldots, m-1$.

Theorem 4.2. *If $n < \infty$ and condition* (c-c) *holds then there exist a constant λ and sequences $\mathbf{x}^* = (x_1^*, \ldots, x_n^*)$, $\mathbf{y}^* = (y_1^*, \ldots, y_n^*)$ with positive terms such that for each solution $\mathbf{x}(t)$ we have*

$$\lim_{t \to \infty} e^{-\lambda t} \mathbf{x}(t) = \mathbf{x}^* \langle \mathbf{y}^*, \mathbf{x}(0) \rangle . \tag{4.17}$$

For the proof see [28] Theorem A.45. The formula (4.17) can be written in the following way

$$\mathbf{x}(t) \approx C e^{\lambda t} \mathbf{x}^*$$

and also in this case we say that the population has an *exponential asynchronous growth.*

The condition (c-c) is weaker than (d-c) one. It is sufficient to check that all subpopulations are connected by oriented paths. For example, the system

$$\begin{cases} x_1' = b_1 x_n - a_1 x_1, \\ x_i' = b_i x_{i-1} - a_i x_i, \quad \text{for } i = 2, \ldots, n, \end{cases} \tag{4.18}$$

where $b_i > 0$ for $1 \leq i \leq n$, satisfies this condition.

Now, we present models with infinite number of subpopulations.

Example 4.6 (birth-death process). *We consider a population of cells. This population is divided into subpopulations in such a way that the i-th subpopulation, $i \geq 0$, consists of cells which contains i copies of a given gen. The length of live of a cell of the type i has exponential distribution with expected value $1/\lambda_i$. Cells of the type i can mutate in the time interval $(t, t + \Delta t)$ to the type $i+1$ with probability $b_i \Delta t + o(t)$ and to the type $i-1$ with probability $d_i \Delta t + o(t)$. Let $x_i(t)$ be the number of cells in the i-th subpopulation. Then*

$$x_i'(t) = -a_i x_i(t) + b_{i-1} x_{i-1}(t) + d_{i+1} x_{i+1}(t), \quad i \geq 0,$$

where $a_i = \lambda_i + b_i + d_i$ *and* $b_{-1} = 0$, $d_0 = 0$.

Example 4.7 (paralog families). *Now we present a model of the evolution of paralog families in a genome [45]. Two genes present in the same genome are said to be* paralogs *if they are genetically identical. It is not a precise definition of paralogs but it is sufficient for our purposes. We are interested in the size distribution of paralogous gene families in a genome. We divide genes into classes. The* i*-th class consists of all* i*-element paralog families. Let* x_i *be a number of families in the* i*-th class. Basing on experimental data Słonimski et al. [46] suggested that*

$$x_i \sim \frac{1}{2^i i}, \quad i = 2, 3, \ldots,$$

but Huynen and van Nimwegen [47] claimed that

$$x_i \sim i^{-\alpha}, \quad i = 1, 2, 3, \ldots,$$

where $\alpha \in (2, 3)$ *decreases if the total number of genes increases. It is very difficult to decide which formula is correct if we study only experimental data because we can only compare first few elements of both sequences. We construct a simple model of the evolution of paralog families which can help to solve this problem.*

The model is based on three fundamental evolutionary events: gene loss, duplication and accumulated change called for simplicity mutation. A single gene during time interval of length Δt *can be:*

- duplicated *with probability* $d\Delta t + o(\Delta t)$ *and duplication of it in a family of the* i*-th class moves this family to the* $(i+1)$*-the class,*
- removed *from the genome with probability* $r\Delta t + o(\Delta t)$. *For* $i > 1$, *removal of a gene from a family of the* i*-th class moves this family to the* $(i-1)$*-th class; removal of a gene from one-element family results in elimination of this family from the genome. A removed gene is eliminated permanently from the pool of all genes.*
- changed *with probability* $m\Delta t + o(\Delta t)$ *and the gene starts a new one-element family and it is removed from the family to which it belonged.*

It is assumed that $\lim_{\Delta t \to 0} \frac{o(\Delta t)}{\Delta t} = 0$. *Moreover, we assume that all elementary events are independent of each other. Let* $s_i(t)$ *be the number of* i*-element families in our model at the time* t. *It follows from the description* *of our model that*

$$s_1'(t) = -(d+r)s_1(t) + 2(2m+r)s_2(t) + m\sum_{k=3}^{\infty} ks_k(t), \tag{4.19}$$

$$s_i'(t) = d(i-1)s_{i-1}(t) - (d+r+m)is_i(t) + (r+m)(i+1)s_{i+1}(t) \tag{4.20}$$

for $i \geq 2$. Let $s(t) = \sum_{i=1}^{\infty} s_i(t)$ be the total number of families. Then the sequence $(p_i(t))$, where $p_i(t) = s_i(t)/s(t)$ is the size distribution of paralogous gene families in a genome at time t.

In order to study properties of system (4.19)–(4.20) we need to introduce some mathematical notions. Let X be the space of sequences (x_i) which satisfies the condition $\sum_{i=1}^{\infty} i|x_i| < \infty$. The main result of the paper [45] is the following.

Theorem 4.3. *There exists a sequence $(s_i^*) \in X$ such that for every solution $(s_i(t))$ of (4.19) and (4.20) with $(s_i(0)) \in X$ we have*

$$\lim_{t\to\infty} e^{(r-d)t} s_i(t) = Cs_i^* \tag{4.21}$$

for every $i = 1, 2, \dots$ and C dependent only on the sequence $(s_i(0))$. Moreover if $d = r$ then

$$\lim_{t\to\infty} s_i(t) = C\frac{\alpha^i}{i}, \tag{4.22}$$

where $\alpha = \dfrac{r}{r+m}$.

In the case when $d = r$ the total number of genes in a genome is constant. It means that the genome is in a stable state. In this case the distribution of paralog families is similar to that stated in Słonimski's conjecture, both distribution are the same if $r = d = m$.

Now we give a sketch of the proof of Theorem 4.3. First, we notice that since the system (4.19) and (4.20) contains infinite number of equations we should formally define its solutions and prove their existence and uniqueness. Moreover we cannot apply the Perron theorem to show the exponential asynchronous growth of the population. Instead of it we apply the lower function theorem of Lasota and Yorke to prove this result. First, we change variables. Let

$$y_i(t) = e^{(r-d)t} is_i(t). \tag{4.23}$$

Then

$$y_1' = -(2d+m)y_1 + (m+r)y_2 + \sum_{k=1}^{\infty} my_k, \tag{4.24}$$

$$y_i' = -(d+r+m+\tfrac{d-r}{i})iy_i + diy_{i-1} + (r+m)iy_{i+1} \tag{4.25}$$

for $i \geq 2$. The system (4.24) and (4.25) generates a stochastic semigroup on l^1, where l^1 denote the space of absolutely summable sequences. We recall that a linear mapping $P : l^1 \to l^1$ is called a *stochastic* or *Markov operator* if $P(D) \subset D$, where

$$D = \left\{ x \in l^1 : x_i \geq 0 \text{ for all } i \geq 1 \text{ and } \sum_{i=1}^{\infty} x_i = 1 \right\}.$$

A family $\{P(t)\}_{t\geq 0}$ of stochastic operators which satisfies conditions:

(a) $P(0) = \mathrm{Id}$,
(b) $P(t+s) = P(t)P(s)$ for $s, t \geq 0$,
(c) for each $x \in l^1$ the function $t \mapsto P(t)x$ is continuous with respect to the l^1 norm

is called a *stochastic* or *Markov semigroup*. The system (4.24) and (4.25) can be written in the following way:

$$y'(t) = Qy(t), \tag{4.26}$$

where $Q = (q_{i,j})_{i,j\geq 1}$. The matrix Q is a *Kolmogorov* matrix. It means that it has the following properties:

(i) $q_{i,j} \geq 0$ for $i \neq j$,
(ii) $\sum_{i=1}^{\infty} q_{i,j} = 0$ for $j \geq 1$.

We also denote by Q the operator $x \mapsto Qx$ with the domain $D(Q) = \{x \in l^1 : Qx \in l^1\}$. The operator Q generates a stochastic semigroup. This result can be obtained from the following theorem.

Theorem 4.4. *Let the matrix Q satisfies conditions (i) and (ii). Let $Q^* = (q^*_{i,j})_{i,j\geq 1}$, where $q^*_{i,j} = q_{j,i}$ for $i, j \geq 1$ and let θ be a positive constant. Then the operator Q generates a stochastic semigroup $\{P(t)\}_{t\geq 0}$ on l^1 if and only if there is no non-zero solution of the equation $Q^*x = \theta x$, where $x \in l^\infty$.*

Recall that l^∞ is the space of bounded sequences. The proof of Theorem 4.4 can be found in [48] Corollary 2.7.3, [49] Proposition 8.3.22, [50] Theorem 23.12.6.

The semigroup $\{P(t)\}_{t\geq 0}$ is called a *semigroup generated* by Eq. (4.26). We also say in this case that the operator (matrix) Q generates the semigroup $\{P(t)\}_{t\geq 0}$. If $y^0 \in l^1$, then the function $y(t) = P(t)y^0$ is called the solution of (4.26) with the initial condition $y(0) = y^0$. The substitution (4.23) defines us the solution of the system (4.19)–(4.20).

The formula (4.21) can be obtained using some result concerning asymptotic stability of stochastic semigroups. A stochastic semigroup is called *asymptotically stable* if there exists $x^* \in D$ with $P(t)x^* = x^*$ for $t > 0$ and such that for every $x \in D$ $\lim_{t\to\infty} \|P(t)x - x^*\| = 0$. Our semigroup is asymptotically stable and the proof of this result is based on the following Lasota-Yorke theorem [51]:

Theorem 4.5. *Let $\{P(t)\}_{t\geq 0}$ be a stochastic semigroup on l^1. If there exists $h \in l^1$, $h \geq 0$ and $h \neq 0$ such that*

$$\lim_{t\to\infty} \|(P(t)x - h)^-\| = 0 \tag{4.27}$$

for every $x \in D$, then the semigroup is asymptotically stable.

Here we use the notation $x_i^- = 0$ if $x_i \geq 0$ and $x_i^- = -x_i$ if $x_i < 0$. The proof of asymptotic stability of our semigroup by using the Lasota-Yorke theorem is very simple so we give it here. Let $y(0) \in D$ and $m > 0$. Then $y(t) \in D$ and since $\sum_{i=1}^{\infty} y_i(t) = 1$ from (4.24) it follows that

$$y_1'(t) \geq -(2d + m)y_1(t) + m.$$

This implies that

$$\liminf_{t\to\infty} y_1(t) \geq \frac{m}{2d+m}.$$

Let $h = (\frac{m}{2d+m}, 0, 0, \ldots)$. Then h satisfies (4.27). From Theorem 4.5 it follows that the semigroup generated by the system (4.24) and (4.25) is asymptotically stable.

Remark 4.1. In the next two part of our lectures we will use stochastic operators and semigroups defined on the space $L^1 = L^1(X, \Sigma, m)$, where X is any set, Σ is a σ-algebra of subsets of X and m is a measure defined on Σ. A linear mapping $P : L^1 \to L^1$, is called a *stochastic* or *Markov operator* if $P(D) \subset D$, where D is the set of densities, *i.e.*

$$D = \{f \in L^1 : f \geq 0 \quad \text{and} \quad \int_X f(x)\, m(dx) = 1\}.$$

We define a stochastic semigroup on L^1 and asymptotic stability as in the case l^1. The Lasota-Yorke lower function theorem remains true in L^1.

Theorem 4.2 does not hold for positive semigroups on the space l^1. But for stochastic semigroups we have the following result.

Theorem 4.6. *Let $\{P(t)\}_{t\geq 0}$ be a stochastic semigroup on l^1 generated by the matrix Q. If condition* (c-c) *holds then w have the following alternative:*
(a) if the semigroup $\{P(t)\}_{t\geq 0}$ have an invariant density, then it is asymptotically stable,
(b) if the semigroup $\{P(t)\}_{t\geq 0}$ have no invariant density, then for each $x \in l^1$ and $i \in \mathbb{N}$ we have

$$\lim_{t\to\infty} (P(t)x)_i = 0. \tag{4.28}$$

Theorem 4.6 follows from some general results of the theory of Markov operators, namely, part (a) from [52] Theorem 2 and part (b) from [53] Theorem 2.

The property (b) is called *sweeping.* Theorem 4.6 can be a useful tool to study the long time behavior of stochastic semigroups. Usually, it is easy to check condition (c-c) and sometimes it is also easy to find an invariant density x^* or check that it does not exist, because it is a solution of the equation $Qx^* = 0$.

Example 4.8 (birth-death process – asymptotics). *We consider the birth-death process given by the equation*

$$x_i'(t) = -a_i x_i(t) + b_{i-1}x_{i-1}(t) + d_{i+1}x_{i+1}(t), \quad i \geq 0,$$

where $a_i = b_i + d_i$ and $b_{-1} = 0$, $d_0 = 0$. The matrix Q corresponding to this equation is a Kolmogorov matrix. If we assume that the birth sequence b_i do not grow too quickly, for example $b_i \leq \alpha i + \beta$ for all i and some α and β, then using Theorem 4.4 one can check that the matrix Q generates a stochastic semigroup. The sequence (x_i^*) which satisfies equation $Qx^* = 0$ is given by the recurrent formula

$$x_{i+1}^* = \frac{b_i + d_i}{d_{i+1}} x_i^* - \frac{b_{i-1}}{d_{i+1}} x_{i-1}^*.$$

If for example $b_i = b$ and $d_i = d$, then one can easily check that

$$x_i^* = c_1 + c_2 \left(\frac{b}{d}\right)^i,$$

where c_1 and c_2 are some constants. It means that the sequence (x_i^*) is a density if and only if $b < d$ and

$$x_i^* = \frac{d-b}{d}\left(\frac{b}{d}\right)^i.$$

4.5. Continuous structure generation models

Now we consider the case when the individual is characterized by one parameter or by a vector of parameters x but time changes in a discrete way. We study the distribution of the population with respect to the parameter x. We want to describe how this distribution changes in consecutive generations of individuals. We begin with a model of the cell-cycle.

Example 4.9 (cell-cycle model). *The cell-cycle is the series of events that take place in a cell leading to its replication. Usually the cell-cycle is divided into four phases. The first one is the growth phase G_1 with the synthesis of various enzymes. Duration of the phase G_1 is highly variable even for cell from one species. DNA synthesis takes place in the second phase S. In the next phase G_2 significant protein synthesis occurs, which is required during the process of mitosis. The last phase M consists of nuclear division and cytoplasmic division. Looking from mathematical point of view we can simplify the model considering only two phases: $A = G_1$ which a random duration t_A and B which consists of the phases S, G_2, and M. The duration t_B of the phase B is almost constant. There are several models of the cell-cycle. Let us mention models by Lasota and Mackey [54] and Tyson and Hannsgen [55]. We present here the Tyrcha model [56] which generalizes the earlier mentioned models.*

In the model of the cell-cycle the crucial role play a parameter x which describes the state of a cell but nobody knows what exactly should be x. It can be size, or contents of genetic material, or amount of *mitogen* (a hypothetical substance responsible for the cell division). We call x the maturity. Let $\varphi(x)$ be the rate of entering the phase B. Let $x(t)$ be the maturity of a cell at time (age) t. The length t_A of the phase A is random and given by formula

$$\text{Prob}(t \le t_A \le t + \Delta t \mid t_A \ge t) \cong \varphi(x(t))\Delta t.$$

Let $g(x)$ be the growth rate, *i.e.* the maturity x grows according to the equation

$$\frac{dx}{dt} = g(x). \tag{4.29}$$

Let $\pi(t, x_0)$ be the size of a cell at time t if its initial size were x_0, *i.e.* $\pi(t, x_0)$ is the solution of (4.29) at time t if it started from x_0. Let $F(t) =$

$\text{Prob}(t_A \geq t)$. Then

$$\frac{F(t) - F(t + \Delta t)}{F(t)} \cong \varphi(\pi(t, x_0))\Delta t.$$

From this equation we obtain

$$F'(t) = -F(t)\varphi(\pi(t, x_0))$$

and after simple calculations we get

$$F(t) = \exp\Big\{ - \int_0^t \varphi(\pi(s, x_0))\, ds \Big\}. \tag{4.30}$$

For $y \geq x_0$ we define $t(x_0, y)$ as such a t that $\pi(t, x_0) = y$. Since

$$\frac{\partial t}{\partial y} \cdot g(\pi(t, x_0)) = 1$$

we receive

$$\frac{\partial t}{\partial y} = \frac{1}{g(y)}$$

and

$$\frac{\partial}{\partial y}\Big(\int_0^{t(x_0,y)} \varphi(\pi(s, x_0))\, ds \Big) = \frac{\varphi(y)}{g(y)}.$$

Let Y be the size of the cell at time t_A. Then

$$\begin{aligned} \text{Prob}(Y \geq y) &= \text{Prob}(\pi(t, x_0) \geq y) = \exp\Big\{ - \int_0^{t(x_0,y)} \varphi(\pi(s, x_0))\, ds \Big\} \\ &= \exp\Big(- \int_{x_0}^{y} \frac{\varphi(r)}{g(r)}\, dr \Big) = \exp(Q(x_0) - Q(y)), \end{aligned}$$

where $Q(y) = \int\limits_0^x \frac{\varphi(r)}{g(r)}\, dr$. Let ξ be a random variable with exponential distribution: $\text{Prob}(\xi \geq x) = e^{-x}$. Then

$$\begin{aligned} \text{Prob}\Big(Q^{-1}(Q(x_0) + \xi) \geq y\Big) &= \text{Prob}\left(Q(x_0) + \xi \geq Q(y)\right) \\ &= \text{Prob}\left(\xi \geq Q(y) - Q(x_0)\right) \\ &= \exp(Q(x_0) - Q(y)) = \text{Prob}(Y \geq y). \end{aligned}$$

From this it follows that the maturity of the cell at the moment of entering the phase B is given by the random variable $Q^{-1}(Q(x_0)+\xi)$ and the initial size of a daughter cell is

$$\gamma\Big(Q^{-1}(Q(x_0) + \xi)\Big),$$

where $\gamma(y) = \frac{1}{2}\pi(t_B, y)$. If X_n is the initial size of cell in the n-th generation then

$$X_{n+1} = \gamma\Big(Q^{-1}\big(Q(X_n) + \xi_n\big)\Big), \tag{4.31}$$

where (ξ_n) is a sequence of independent random variables with exponential distribution. If f_n is the density of the distribution function of X_n then $f_{n+1} = Pf_n$, where

$$Pf(x) = -\int\limits_0^{\lambda(x)} \frac{\partial}{\partial x}\Big\{H\Big(Q(\lambda(x)) - Q(y)\Big)\Big\} f(y)\, dy, \tag{4.32}$$

$H(x) = e^{-x}$ and λ is the inverse function for γ.

The asymptotic properties of the operator P depend on the function $\alpha(x) = Q(\lambda(x)) - Q(x)$. We have

(a) If $\alpha(x) > 1$ for sufficiently large x, then P is *asymptotically stable*, *i.e.* there exists a density f^* such that

$$\lim_{n\to\infty} \|P^n f - f^*\| = 0 \qquad \text{for } f \in D.$$

(b) If $\alpha(x) \le 1$ for sufficiently large x, then P is *sweeping* or *zero type*, *i.e.*

$$\lim_{n\to\infty} \int_0^c P^n f(x)\, dx = 0 \qquad \text{for } f \in D \text{ and } c > 0.$$

(c) If $\inf \alpha(x) > -\infty$, then the operator P is *completely mixing*, *i.e.*

$$\lim_{n\to\infty} \|P^n f - P^n g\| = 0 \qquad \text{for } f, g \in D.$$

The results were proved, respectively, (a) in [57], (b) in [58], and (c) in [59].

Now we go to a general situation. Many biological processes can be modeled by means of randomly perturbed dynamical systems. The relation between the distribution of a parameter (size, maturity, *etc.*) in two successive generations is given by

$$X_{n+1} = S(X_n, \xi_{n+1}), \tag{4.33}$$

where $(\xi_n)_{n=1}^{\infty}$ is a sequence of independent random variables (or elements) with the same distribution, and the initial value of the system X_0 is independent of the sequence $(\xi_n)_{n=1}^{\infty}$. Studying systems of the form (4.33) we are often interested in the behavior of the sequence of the measures (μ_n) defined by

$$\mu_n(A) = \text{Prob}(X_n \in A). \tag{4.34}$$

The evolution of these measures can be described by a Markov operator P given by $\mu_{n+1} = P\mu_n$. The operator P is defined on the space of probability measures. Let m be a given measure in the phase space (*i.e.* in the space of parameters). Assume that the distribution ν_y of the random variable $S(y, \xi_n)$ is absolutely continuous with respect to m and let $k(x, y)$ be the density of ν_y. Assume that the measure μ_n is absolutely continuous with respect to the measure m and let f_n be the density $\dfrac{d\mu_n}{dm}$. Then the measure μ_{n+1} has a density $f_{n+1} = Pf_n$, where the operator P is given by the formula

$$Pf(x) = \int k(x, y)f(y)\, m(dy).$$

4.6. Continuous time-structure models

In this section we present models in which both the time and the structure is continuous. In such models an individual is described by a parameter $x \in \mathbb{R}^n$ (age, size, maturity etc.). We are interested in finding the distribution of the parameter x at time t. This distribution is described by a density function $u(t, x)$, exactly

$$\int_A u(t, x)\, dx$$

is the number of individuals (or biomass) with the parameter x in set A. It should be underline that u is not a density in the probabilistic sense because the integral of $u(t, x)$ over the whole phase space is not one and this integral can change with time.

It is a large class of various models of this type and it is rather difficult to give a unified description of them. We begin with some examples and then we will try to present a general approach to these models.

Example 4.10 (age-structured model). *In the age structure McKendrick model there is only one parameter – the age of an individual a, which belongs to the interval $[0, c)$, where c a positive number or infinity. We assume that in the time interval $[t, t + \Delta t]$ an individual with age a can*

(a) with probability $b(t, a)\Delta t + o(\Delta t)$ "produce" a new individual. This assumption leads to following integral equation

$$u(t, 0) = \int_0^c b(t, a)u(t, a)\, da,$$

(b) with probability $\mu(t,a)\Delta t + o(\Delta t)$ die.

From condition (b) it follows that

$$u(t+\Delta t, a+\Delta t) - u(t,a) = -\mu(t,a)u(t,a)\Delta t + o(\Delta t).$$

If we pass with Δt to 0 we obtain a partial differential equation which is written below. We should also add an initial condition and, as a result, the whole model consists of three equations:

$$\frac{\partial u}{\partial t} + \frac{\partial u}{\partial a} = -\mu(t,a)u, \tag{4.35}$$

$$u(t,0) = \int_0^c b(t,a)u(t,a)\,da, \tag{4.36}$$

$$u(0,a) = u_0(a). \tag{4.37}$$

Let $N(t) = \int_0^c u(t,a)\,da$ be the total number of individuals at time t. Then the function $p(t,a) = \frac{u(t,a)}{N(t)}$ is the probability density of the age distribution at time t. The function $p(t,a)$ is called the *age profile.*

The following theorem describes the long time behavior of the solutions of (4.35). We consider the case $c < \infty$. We precede the formulation of the theorem by some condition. We will assume that the initial function $u(0,a)$ satisfies the following condition:

$$\int_0^c \int_0^{c-a} u(0,a+t)b(t,a+t)\,dt\,da > 0. \tag{4.38}$$

Condition (4.38) denotes that the initial distribution of the population is chosen in such a way that not everybody is beyond child-bearing age.

Theorem 4.7 (Ergodicity, Norton(1928)). *Assume that μ and b are continuous functions and that there exist $a_0 \in (0,c)$, $\varepsilon > 0$ and $\delta > 0$ such that $b(t,a) > \varepsilon$ for $a \in (a_0-\delta, a_0+\delta)$ and $t \geq 0$. Let u, $\bar{u}$ be the solutions with the initial conditions $u(0,a)$, $\bar{u}(0,a)$ satisfying* (4.38). *Then*

$$\lim_{t\to\infty} \frac{\bar{p}(t,a)}{p(t,a)} = 1.$$

In the formulation of Theorem 4.7 the birth and death rates can depend on time and profiles can also depend on time. More precise information about long time behavior can be obtained if we assume that the birth and death rates do not depend on time or they are periodic functions (see corollaries below).

Corollary 4.1 (Asynchronous exponential growth). *If b and d are independent on t, then there exist a constant $\lambda \in \mathbb{R}$ and a function $p_*(a)$ independent of a solution u and a constant C dependent on u such that*

$$\lim_{t\to\infty} e^{-\lambda t} N(t) = C \quad \textit{and} \quad \lim_{t\to\infty} p(t,a) = p_*(a).$$

Corollary 4.2 (Periodic case). *If b and d are periodic functions of t with period T, then there exist a constant $\lambda \in \mathbb{R}$ and a function $p_*(t,a)$ periodic with respect to t with period T such that for each solution u there exists a constant C which satisfy*

$$\lim_{t\to\infty} \frac{u(t,a)}{e^{\lambda t} p_*(t,a)} = C.$$

Theorem 4.7 and Corollaries 4.1 and 4.2 were generalized in [60], [61], [62]. In these papers it is considered a population divided into n subpopulations. These subpopulations can be different phenotypes, or demographically distinct populations, or with $n = 2$ a two sex population, etc. We assume that for each i and j an individual from the i-th subpopulation can have descendants in the j-th subpopulation and consider an age-structured model. Let $u_i(t,a)$ describes the age distribution in the i-th subpopulation. Then the functions $u_i(t,a)$ satisfy a system of partial differential equations of the first order with boundary-initial conditions. In this case one can prove that for two solutions u and $\bar{u}$ there exists a constant $c > 0$ such that

$$\lim_{t\to\infty} \frac{\bar{u}_i(t,a)}{u_i(t,a)} = c.$$

Moreover, if the rates of death, birth and transition between subpopulations are periodic with the same period T or do not depend on time then Corollaries 4.1 and 4.2 also hold.

Let us return back to the McKendrick model. Assume that the birth and death rates do not depend on time, *i.e.* $\mu(t,a) = \mu(a)$ and $b(t,a) = b(a)$. We want to find the growth rate and the age profile for large time. If $u(t,a) = e^{\lambda t} p(a)$ is the solution of the McKendrick system then

$$\lambda p(a) + p'(a) = -\mu(a)p(a), \tag{4.39}$$

$$p(0) = \int_0^c b(a)p(a)\,da. \tag{4.40}$$

From (4.39) we get

$$\begin{aligned} p(a) &= p(0)\exp\Big\{-\int_0^a (\lambda+\mu(s))\,ds\Big\} \\ &= p(0)e^{-\lambda a}\exp\Big\{-\int_0^a \mu(s)\,ds\Big\}. \end{aligned} \tag{4.41}$$

Let

$$\varphi(\lambda) = \int_0^c b(a)e^{-\lambda a}\exp\Big\{-\int_0^a \mu(s)\,ds\Big\}\,da. \tag{4.42}$$

Then condition (4.40) holds if and only if $\varphi(\lambda)=1$. It is easy to check that $\varphi(-\infty)=+\infty$, $\varphi(+\infty)=0$ and $\varphi'(\lambda)<0$ for all $\lambda\in\mathbb{R}$. This implies that there exists a unique $\lambda_0\in\mathbb{R}$ such that $\varphi(\lambda_0)=1$. Since

$$p(a) = p(0)e^{-\lambda_0 a-\int_0^a \mu(s)\,ds}.$$

and since the function $p(a)$ is a probability density, *i.e.* $\int_0^c p(a)\,da=1$, we have

$$p(a) = e^{-\lambda_0 a-\int_0^a \mu(s)\,ds}\Big(\int_0^c e^{-\lambda_0 a-\int_0^a \mu(s)\,ds}\,da\Big)^{-1}.$$

Let T be the length of the life of an individual. Then T is a random variable. The function $F(a)=P(T\ge a)$ is the probability of being still alive at age a and this function is called the *survival function.* Now, we find the survival function. Since

$$P(T\in[a,a+\Delta a)\,|\,T\ge a) = \mu(a)\Delta a + o(\Delta a)$$

we have

$$\frac{F(a)-F(a+\Delta a)}{F(a)} = \mu(a)\Delta a + o(\Delta a)$$

and, consequently,

$$\frac{F'(a)}{F(a)} = -\mu(a).$$

It means that

$$F(a) = \exp\Big(-\int_0^a \mu(s)\,ds\Big)$$

and we can write the age profile $p(a)$ in the following way

$$p(a) = e^{-\lambda_0 a}F(a)\Big(\int_0^c e^{-\lambda_0 a}F(a)\,da\Big)^{-1}.$$

Now, we return back to the beginning of our lectures, namely, to Ulpian's table. Let us assume that we are able to collect data concerning the distribution of the duration of life in a population. Having these data we are able to calculate the survival function F. The question is: how to find Ulpian's table? Ulpian's table gives us the information on the expected remaining life. Let $\varphi_a(x)$ be the probability that *remaining life* at age a is at least x. Then

$$\varphi_a(x) = P(T \geq a + x \,|\, T \geq a) = \frac{P(T \geq a + x)}{P(T \geq a)} = \frac{F(a+x)}{F(a)}.$$

It means that the *expected remaining life* at age a is given by the formula

$$D(a) = -\int_0^c x\varphi_a'(x)\,dx = \int_0^c \varphi_a(x)\,dx = \int_0^c \frac{F(a+x)}{F(a)}\,dx.$$

Now we present another age-structured model which is interesting both from biological point of view because it concerns specific but important medical problem, and also from mathematical point of view because the model contains a a nontrivial boundary condition and leads to an interesting delay differential equation.

Example 4.11 (erythrocytes dynamics). *Now we presented a model of the red blood cells dynamics by Ważewska–Czyżewska and Lasota [63]. We start from the necessary biological information. Red blood cells, like other blood cell, are produced in the bone marrow in a process called hematopoiesis from the same cells called committed stem cells in about 7 days. Healthy erythrocytes live about 120 days before they are degraded. The production of them is stimulated by the hormone erythropoietin and the production system try to keep the number of cells on a constant level.*

Now we formulate the model. Let $n(t,a)$ be the age distribution of red cells at time t. Then $N(t) = \int_0^\infty n(t,a)\,da$ is the total number of red cells at time t and $p(t) = n(t,0)$ is the production of new cells in a unit time. Let h be the time for the production of a mature erythrocytes. The degree of arousal of the system can be characterized by the function

$$S(t) = \frac{p'(t)}{p(t)}.$$

We assume that the change of the number of red cells in the blood circulation causes arousal of the system in the following way

$$S(t) = -\frac{d}{dt}\gamma N(t-h).$$

It means that

$$\frac{p'(t)}{p(t)} = -\frac{d}{dt}\gamma N(t-h)$$

and, consequently,

$$p(t) = \rho e^{-\gamma N(t-h)}. \tag{4.43}$$

The constant ρ can be interpreted as the demand of the organism for oxygen. If $\mu(t,a)$ is the rate of degradation of cells then analogously to the McKendrick model we receive the equation

$$\frac{\partial n}{\partial t} + \frac{\partial n}{\partial a} = -\mu(t,a)n. \tag{4.44}$$

Taking into account Eq. (4.43) and the initial condition the whole model is the following

$$\begin{aligned} &\frac{\partial n}{\partial t} + \frac{\partial n}{\partial a} = -\mu(t,a)n, \\ &n(t,0) = \rho e^{-\gamma N(t-h)}, \\ &n(0,a) = n_0(a). \end{aligned}$$

We can also consider a simplified model. Let us assume that

$$\mu = \frac{1}{N(t)} \int_0^\infty \mu(t,a)n(t,a)\,da$$

do not depend on time. The constant μ is called the coefficient of destruction. Integrating both sides of (4.44) over a we get

$$N'(t) + n(t,\infty) - n(t,0) = -\mu N(t)$$

and assuming that $n(t,\infty) = 0$ and using (4.43) we obtain

$$N'(t) = -\mu N(t) + \rho e^{-\gamma N(t-h)}. \tag{4.45}$$

Using computer simulations it is rather easy to find solutions of this equation for different coefficient h, γ, and ρ and compare the results with empirical data. The paper [63] contains conclusions concerning the course of the diseases under different parameters and the medical treatment of them. Since for some parameters Eq. (4.44) has periodic solutions the model confirmed the existence of periodic blood diseases and that this periodicity is a property of the system and does not follow from the daily or another typical rhythm of life [64].

Example 4.12 (size structured model). *Now we consider a model of a cellular population which was introduced for the first time probably by Bell and Anderson [65] and was studied and generalized in many papers (see e.g. [66], [67], [6], [68], [69]). In this model a cell is characterized by its size x, a number from the interval $[a,1]$, $0<a<1$. We assume that the death and the division rates are $d(x)$ and $b(x)$, respectively. We also assume that if x is the size of the mother cell then after division the size of the daughter cells is $x/2$. It is obvious that $b(x)=0$ for $x<2a$ and we need to assume that*

$$\int_a^1 b(x)\,dx=\infty.$$

This condition guarantees us that the size of any cell cannot be greater than 1. Observe that the lost of cells with size $\le m$ in the time interval of the length Δt is

$$\Delta t\int_0^m (d(x)+b(x))u(t,x)\,dx+o(\Delta t)$$

and the number of the new cells with size $\le m$ in this time interval is given by

$$\Delta t\int_0^{2m} 2b(r)u(t,r)\,dr+o(\Delta t).$$

Moreover we assume that a cell grows according to the equation $x'=g(x)$. It means that the cells with size $\le m$ at time t will have size $\le m+g(m)\Delta t+o(\Delta t)$ at time $t+\Delta t$. Combining these three elements we obtain

$$\begin{aligned}\int_0^{m+g(m)\Delta t} u(t+\Delta t,x)\,dx-\int_0^m u(t,x)\,dx\\ =-\Delta t\int_0^m (d(x)+b(x))u(t,x)\,dx\\ +\Delta t\int_0^{2m} 2b(r)u(t,r)\,dr+o(\Delta t).\end{aligned}\tag{4.46}$$

Dividing both sides of (4.46) by Δt and letting $\Delta t\to 0$ we receive

$$\begin{aligned}\int_0^m \frac{\partial}{\partial t}u(t,x)\,dx+g(m)u(t,m)=-\int_0^m (d(x)+b(x))u(t,x)\,dx\\ +\int_0^{2m} 2b(r)u(t,r)\,dr.\end{aligned}$$

Differentiating both sides of the last equation with respect to m we finally obtain

$$\frac{\partial}{\partial t}u(t,x)+\frac{\partial}{\partial x}(g(x)u(t,x)) = -(d(x)+b(x))u(t,x)+4b(2x)u(t,2x). \quad (4.47)$$

The full model consists of Eq. (4.47) and the following boundary and initial conditions

$$u(t,a) = 0, \quad (4.48)$$

$$u(0,x) = v(x). \quad (4.49)$$

Examples 4.10 and 4.12 show that though both models are based on similar biological assumptions they lead to different mathematical objects. Our aim is to give a unified approach to structured models. We start with the *continuity equation.* Consider a structured model without mortality and proliferation. Let $x \in G \subset \mathbb{R}^n$ be a parameter which characterizes any individual. We assume that the parameter x changes according to the equation

$$x'(t) = g(t, x(t)). \quad (4.50)$$

If $u(t,x)$ is the distribution of x then u satisfies the following equation

$$\frac{\partial u(t,x)}{\partial t} + \operatorname{div}(g(t,x)u(t,x)) = 0, \quad (4.51)$$

where

$$\operatorname{div}(g(t,x)u(t,x)) = \sum_{i=1}^{n} \frac{\partial}{\partial x_i}(g(t,x)u(t,x)). \quad (4.52)$$

Proof. Given a domain $D \subset G$ with the smooth boundary S, consider the fluxes into the set D in the time interval of the length Δt:

$$I(\Delta t) = \int_D u(t+\Delta t, x)\, dx - \int_D u(t,x)\, dx. \quad (4.53)$$

Since the fluxes are through the surface S and since the speed at which individuals cross the surface is $-n(x) \cdot g(t,x)$, where $n(x)$ is the outward-pointing unit normal vector to S, we have

$$I(\Delta t) = -\Delta t \int_S (n(x) \cdot g(t,x)u(t,x))\, d\sigma(x) + o(\Delta t). \quad (4.54)$$

According to the Gauss-Ostrogradski theorem we have

$$\int_S (n(x) \cdot g(t,x)u(t,x))\, d\sigma(x) = \int_D \operatorname{div}(g(t,x)u(t,x))\, dx. \quad (4.55)$$

Equations (4.53), (4.54) and (4.55) imply (4.51). □

Now we introduce the *general reproduction operator*. We assume that an individual with the parameter x has k descendants and that $\mathcal{P}_k(x, A)$ is the probability that any of its descendant has the parameter in the set $A \subset G$ at the birth. For example, if x is the age then $\mathcal{P}_k(x, A) = \mathbf{1}_A(0)$. If x is the size then

$$\mathcal{P}_2(x, A) = \begin{cases} 1, & \text{if} \quad x/2 \in A, \\ 0, & \text{if} \quad x/2 \notin A. \end{cases}$$

Let $b_k(x)\Delta t$ be the probability that an individual with parameter x has k descendants in time interval $[t, t + \Delta t]$. We set

$$\mathcal{P}(x, A) = \sum_{k=1}^{\infty} k b_k(x) \mathcal{P}_k(x, A).$$

Then $\mathcal{P}(x, A)\Delta t$ is the probability that an individual with parameter x has a descendant in the set A. We also assume that $\mu(t, x)$ is the death rate which includes both the real death and the lost of cells during the process of division.

Now we introduce the *Kolmogorov's backward equation* corresponding to our model. Denote by $m_{t,x}$ the measure which describes the distribution of the parameter x at time t if at the initial time 0 we have one individual with parameter x. Let

$$u(t, x) = \int_G f(y) m_{t,x}(dy)$$

for a smooth function g. Then the function u satisfies the Kolmogorov's backward equation

$$\frac{\partial u}{\partial t} = \underbrace{-\mu u + \sum_{i=1}^{n} g_i \frac{\partial u}{\partial x_i} + \int_G u(t, y)\, \mathcal{P}(x, dy)}_{\mathcal{A}^* u} \tag{4.56}$$

and $u(0, x) = f(x)$. Set

$$Sf(x) = \int_G f(y)\, \mathcal{P}(x, dy).$$

Then the Eq. (4.56) can be written in the following way

$$\frac{\partial u}{\partial t} = -\mu u + \sum_{i=1}^{n} g_i \frac{\partial u}{\partial x_i} + Su. \tag{4.57}$$

The equation describing the evolution of the densities of the distributions of the parameter x in the population is conjugated to Eq. (4.56) and is called the *Kolmogorov's forward (Fokker-Planck) equation*:

$$\frac{\partial u}{\partial t} = \mathcal{A}u. \tag{4.58}$$

But the question is: does there exist a linear operator $\mathcal{A}$ defined on a dense subspace of $L^1(G)$ such that $\mathcal{A}^*$ is given by formula (4.56) ?

Example 4.13. If there exists a linear bounded operator $P : L^1(G) \to L^1(G)$ such that $P^* = S$, then

$$\mathcal{A}f = -\mu f - \operatorname{div}(gf) + Pf.$$

The sufficient condition for the existence of the operator P is the following:
(C) *if* $l(B) = 0$, *then* $\mathcal{P}(x, B) = 0$ *for almost all* x,
where l denotes the Lebesgue measure and B is any Borel set. Indeed, from (C) it follows that for each density f the measure m given by the formula

$$m(B) = \int_G f(x)\mathcal{P}(x, B)\,dx, \quad \text{for Borel sets } B,$$

is absolutely continuous with respect to the Lebesgue measure and $m(G) = 1$. From this it follows that the Radon-Nikodym derivative $\frac{dm}{dx}$ exists and it is a density. The operator P is given by the formula $Pf = \frac{dm}{dx}$. For example, in the size structured population model we have

$$\mathcal{P}(x, A) = \begin{cases} 2b(x), & \text{if } \ x/2 \in A, \\ 0, & \text{if } \ x/2 \notin A. \end{cases}$$

Then the transition function $\mathcal{P}$ satisfies (C) and the operator P is given by the formula $Pf(x) = 4b(2x)f(2x)$.

Example 4.14. In the age structure McKendrick model the transition function is the following $\mathcal{P}(x, A) = 2b(x)\mathbf{1}_A(0)$. This transition function does not fulfill condition (C) and the operator P does not exist. But in this case we can assume that

$$\mathcal{A}f(a) = -\mu(a)f(a) - f'(a)$$

and the domain of the operator $\mathcal{A}$ is the following

$$D(\mathcal{A}) = \{f \in L^1 : f' \in L^1, \ f(0) = \int_0^c 2b(a)f(a)\,da\}.$$

Simple calculations show that

$$\mathcal{A}^* f(a) = -\mu(a)f(a) + f'(a) + 2b(a)f(0).$$

It means that

$$Sf(a) = 2b(a)f(0) = \int f(x)\mathcal{P}(a, dx).$$

Thus we obtain the proper Kolmogorov's backward equation:

$$\frac{\partial}{\partial t}u(t,a) = -\mu(a)u(t,a) + \frac{\partial}{\partial a}u(t,a) + 2b(a)u(t,0).$$

In general, the long time behavior of linear continuous structured models is similar to that in the McKendrick model – they usually have asynchronous exponential growth. We illustrate it considering the size structured model.

Theorem 4.8. *If $g(2x) \neq 2g(x)$ at least for one $x \in [a,1]$, then there exist $\lambda \in \mathbf{R}$ and positive functions f_* and w such that*

$$e^{-\lambda t}u(t,\cdot) \to f_* \int_a^1 u(0,x)w(x)\,dx \quad \text{in} \quad L^1(a,1).$$

Sketch of the proof. We repeat some steps in the proof of a more general result from the paper [68]. Equation (4.47) can be written as an evolution equation $u'(t) = Au$. First we show that A is an infinitesimal generator of a continuous semigroup $\{T(t)\}_{t\geq 0}$ of linear operators on $L^1(a,1)$. Then we prove that there exist $\lambda \in \mathbb{R}$ and continuous and positive functions v and w such that $Av = \lambda v$ and $A^*w = \lambda w$. From this it follows that the semigroup $\{P(t)\}_{t\geq 0}$ given by $P(t) = e^{-\lambda t}T(t)$ is a stochastic semigroup on the space $L^1(X,\Sigma,m)$, where m is a Borel measure on the interval $[a,1]$ given by $m(B) = \int_B w(x)\,dx$. Moreover, for some $c > 0$ the function $f_* = cv$ is an invariant density with respect to $\{P(t)\}_{t\geq 0}$. Finally, from Theorem 4.9 (see below) we conclude that this semigroup is asymptotically stable. Since the Lebesgue measure and the measure m are equivalent we obtain that $e^{-\lambda t}u(t,\cdot)$ converges to $f_*\Phi(u(0,\cdot))$ in $L^1(a,1)$, where $\Phi(g) := \int_a^1 g(x)w(x)\,dx$. □

In the proof of Theorem 4.8 we use some general result concerning stochastic semigroups. We recall it now. A stochastic semigroup $\{P(t)\}_{t\geq 0}$ is called *partially integral* if there exist $t_0 > 0$ and a measurable nonnegative function $q(x,y)$ such that

$$\int_X \int_X q(x,y)\,m(dx)\,m(dy) > 0 \tag{4.59}$$

and

$$P(t_0)f(x) \geq \int_X q(x,y)f(y)\, m(dy) \quad \text{for every } f \in D. \tag{4.60}$$

Theorem 4.9 ([52]). *Let $\{P(t)\}_{t\geq 0}$ be a partially integral stochastic semigroup. Assume that the semigroup $\{P(t)\}_{t\geq 0}$ has the only one invariant density f_*. If $f_* > 0$ a.e., then the semigroup $\{P(t)\}_{t\geq 0}$ is asymptotically stable.*

The general scheme described above includes a large family of structured models. The parameters can be also the space position, maturity, contents of genetic material etc., and we can consider models which includes more parameters: *e.g.* age-size, space-maturity structured models. But there are a lot of models which are outside of this scheme. Now we give some examples.

Example 4.15 (simple cell-cycle model). *We consider a model of the cell-cycle given by Rubinow [70]. In this model the state of a cell in the cell-cycle is described by a parameter called maturity $0 \leq x \leq 1$. It is assumed that the parameter x grows according to the equation*

$$x' = g(x)$$

and the cell divides at maturity 1 and new born cells have maturity 0. Let $u(t,x)$ be the density of the population with respect to x. We also assume that the mortality rate is $\mu(x)$. The density $u(t,x)$ satisfies the following initial-boundary problem

$$\begin{aligned}
&\frac{\partial}{\partial t}u(t,x) + \frac{\partial}{\partial x}(g(x)u(t,x)) = -\mu(x)u(t,x),\\
&g(0)u(t,0) = 2g(1)u(t,1),\\
&u(0,x) = u_0(x).
\end{aligned}$$

Although the model is similar to the age-structured model it cannot be treated by our general scheme because we cannot find the proper transition function $\mathcal{P}$ for it.

Example 4.16 (continuous Penna model). *Now we present a continuous version of the Penna model [71]. In this model both time and age variables are nonnegative real numbers. Each individual is described by its age and its own maximum life span m. Let $u(t,a,m)$ be the density of a* *and m at time t. Then the model is described by the system of equations:*

$$\begin{aligned} N(t) &= \int_0^\infty \int_0^m u(t,a,m), \\ \frac{\partial u}{\partial t} + \frac{\partial u}{\partial a} &= -\mu(a,m,N(t))u(t,a,m), \quad \text{for } a < m, \\ u(t,a,m) &= 0, \quad \text{for } a \geq m, \\ u(t,0,m) &= \int_0^\infty \int_0^{m'} b(a,m,m',N(t))u(t,a,m')\,da\,dm'. \end{aligned} \tag{4.61}$$

As in Example 4.5, μ is the death rate. If the size of the population is N then the probability that a parent with maximum life span m' and age a gives birth to a child with maximum life span $m \in (m_1, m_2)$ in the time interval of the length Δt is

$$\Delta t \int_{m_1}^{m_2} b(a,m,m',N)\,dm.$$

In [71] it is assumed that like in the Penna model we have $\mu = N(t)/N_{\max})$ and $b = (1 - N(t)/N_{\max})p(m,m')$, where $N_{\max}$ is the maximum size of the population and $p(m,m')$ is the birth function. The operator $Pf(m) = \int_0^\infty f(m')\,dm'$ can be called the mutation operator.

In some models we consider both continuous and discrete parameters. For example in tumor growth models we consider cells in a few different states connected with malignant progression but we are interested in the space distribution of them. A similar situation appears in epidemiology if we consider models with the space or age distribution of susceptible, infected, and resistant people [5]. In such a situation we consider different densities $u_1,\ldots,u_n$ for different classes of individuals. Now we present another model of this type.

Example 4.17 (the age-structured model with telomere loss).
In [72] the authors consider an age-structured cell population model in which the population is split into $N+1$ subpopulations according to the telomere state. The biological aspects of telomere shortening were discussed in Example 4.3. We are interested in the age distribution $u_i(t,a)$ of cells in the i-th subpopulation, $i = 0, 1, \ldots, N$. The functions $u_0, \ldots, u_N$ satisfies the following system of $(N+1)$ partial differential equations with $(N+1)$

boundary and initials conditions:

$$\frac{\partial}{\partial t}u_j(t,a) + \frac{\partial}{\partial a}u_j(t,a) = -(b_j(a) + \mu_i(a))u_j(t,a), \tag{4.62}$$

$$u_j(t,0) = 2\sum_{k=j}^{N} p_{jk}\int_0^\infty b_k(a)u(t,a)\,da, \tag{4.63}$$

$$u_j(0,a) = v_j(a). \tag{4.64}$$

Here b_i, μ_i are the birth and death rates in the i-th telomere state and p_{jk} represents the probability for a cell in the k-th telomere state to produce by division a cell in the j-th telomere state. We assume that $p_{jk} = 0$ if $k < j$, which means that the daughter cell is in a lower state than the mother cell. As in Example 4.3 we do not have here the asynchronous exponential growth but the population has the asynchronous polynomial exponential growth.

Example 4.18 (stochastic gene expression). *Now we present a simple model of gene expression introduced in the paper by Lipniacki et al. [73] and we give also some analytic results concerning this model received in the paper [74]. We consider the process of the regulation of a single gene. The model involves three processes: allele activation/inactivation, mRNA transcription/decay, and protein translation/decay. A gene can be in an active (denoted by A) or inactive state (denoted by I) and it can be transformed into an active state or into an inactive state, with intensities q_0 and q_1, respectively. The rates q_0 and q_1 depend on the number of mRNA molecules $x_1(t)$ and on the number of protein molecules $x_2(t)$. If the gene is active then mRNA transcript molecules are synthesized at the rate R. The protein translation proceeds with the rate $Kx_1(t)$, where K is a constant. In addition, mRNA and protein molecules undergo the process of degradation. The mRNA and protein degradation rate are m and r, respectively. The reactions described above may be summarized as follows:*

$$I \xrightarrow{q_0} A, \qquad I \xrightarrow{q_1} A, \tag{4.65}$$

$$A \xrightarrow{R} mRNA \xrightarrow{m} \phi, \tag{4.66}$$

and

$$mRNA \xrightarrow{Kx_1(t)} protein \xrightarrow{r} \phi, \tag{4.67}$$

where ϕ stands for degradation of gene products. The state of the system is described by the triple $(x_1(t), x_2(t), \gamma(t))$, where $\gamma(t)$ is a random variable *with value 1 if the gene is in the active state and 0 in the inactive state. The functions $x_1(t)$ and $x_2(t)$ satisfy the following equations*

$$\frac{dx_1}{dt} = H\gamma(t) - mx_1, \tag{4.68}$$

$$\frac{dx_2}{dt} = Kx_1 - rx_2. \tag{4.69}$$

The switching function $\gamma(t)$ is a stochastic process with values in the set $\{0,1\}$ and this process depends on the functions $x_1(t)$ and $x_2(t)$.

Equations (4.68) - (4.69) generate stochastic trajectories, which can be described as piecewise deterministic, time-continuous Markov process

$$p(t) = (x_1(t), x_2(t), \gamma(t)) = (\mathrm{x}(t), \gamma(t)),\ t \geq 0. \tag{4.70}$$

We introduce partial density functions of this process $u_i(t, x_1, x_2)$ by

$$\Pr\{\mathrm{x}(t) \in B, \gamma(t) = i\} = \iint_B u_i(x_1, x_2, t)\, dx_1\, dx_2, \qquad i = 0,1$$

where B is a Borel subset of $\mathbb{R}^+ \times \mathbb{R}^+$. The functions satisfy the following Fokker-Planck system:

$$\begin{aligned} \frac{\partial u_0}{\partial t} + \frac{\partial}{\partial x_1}(-mx_1u_0) + \frac{\partial}{\partial x_2}((Kx_1 - rx_2)u_0) &= q_1u_1 - q_0u_0, \\ \frac{\partial u_1}{\partial t} + \frac{\partial}{\partial x_1}((R - mx_1)u_1) + \frac{\partial}{\partial x_2}((Kx_1 - rx_2)u_1) &= q_0u_0 - q_1u_1, \end{aligned} \tag{4.71}$$

where $q_0 = q_0(x_1, x_2)$ and $q_1 = q_1(x_1, x_2)$ are given non-negative continuous functions defined on the rectangle $\mathcal{K} = [0, R/m] \times [0, KR/Mr]$. Next, we construct a stochastic semigroup $\{P(t)\}_{t\geq 0}$ on the space $L^1(\mathcal{S})$, $\mathcal{S} = \mathcal{K} \times \{0,1\})$, related to the Fokker-Planck system of equations for the densities of the process. The semigroup $\{P(t)\}_{t\geq 0}$ is defined by $P(t)f(x_1, x_2, i) = u_i(t, x_1, x_2)$, where (u_1, u_2) is the solution of the system (4.71) with the initial conditions: $u_0(0, x_1, x_2) = f(x_1, x_2, 0)$ and $u_1(0, x_1, x_2) = f(x_1, x_2, 1)$. The main result of the paper [74] is the asymptotic stability of the semigroup $\{P(t)\}_{t\geq 0}$. The proof of this result is based on the following theorem.

Theorem 4.10. *Let S be a compact metric space and Σ be the σ-algebra of Borel sets. Let $\{P(t)\}_{t\geq 0}$ be a Markov semigroup which satisfies conditions: (a) for every $f \in D$ we have $\int_0^\infty P(t)f\, dt > 0$ a.e.,*

(b) for every $q_0 \in \mathcal{S}$ there exist $\kappa > 0$, $t > 0$, and a measurable function $\eta \geq 0$ such that $\int \eta\, dm > 0$ and

$$P(t)f(p) \geq \eta(p) \int_{B(q_0,\kappa)} f(q)\, m(\mathrm{d}q) \tag{4.72}$$

for $p \in \mathcal{S}$, where $B(q_0, \kappa)$ is the open ball with center q_0 and radius κ, Then the semigroup $\{P(t)\}_{t\geq 0}$ is asymptotically stable.

Theorem 4.10 is a simple consequence of Theorem 4.9 and Theorem 2 [53].

Till now we have discussed rather simple structured models. Now we present a more advanced model [75] to show various problems which can appear in building models and in their investigations.

Example 4.19 (two-phase model of the cell-cycle [75]). *Now we assume like in Example 4.9 that a cell can be in the resting phase A which duration is random or in the proliferating phase which duration is constant τ. The state of a cell is characterized by one parameter m called maturity. Denote by $p(t, m, a)$ and $n(t, m, a)$ the maturity-age distribution of proliferating and resting cells, respectively. The maturity m grows according to the equation*

$$m' = V(m).$$

Let $\gamma(m)$ and $\delta(m)$ denote the death rates in phases A and B. Let $N(t, m) = \int n(t, m, a)\, da$, and $\bar{N}(t) = \int N(t, m)\, dm$. Then $\bar{N}(t)$ is the total number of cells in the resting phase and we can assume that $\bar{N}(t)$ characterizes the state of the whole population. We assume that the rate of entering of the proliferating phase β depends on the maturity of a cell and the state of the whole population, i.e. $\beta = \beta(\bar{N}, m)$. Let $h(m)$ be the maturity of the mother cell at cytokinesis if its daughter cells at birth have the maturity m. The whole model consists of two partial differential equations

$$\frac{\partial p}{\partial t} + \frac{\partial p}{\partial a} + \frac{\partial (Vp)}{\partial m} = -\gamma p, \tag{4.73}$$

$$\frac{\partial n}{\partial t} + \frac{\partial n}{\partial a} + \frac{\partial (Vn)}{\partial m} = -(\delta + \beta)n \tag{4.74}$$

and two boundary conditions

$$p(t, m, 0) = \beta(\bar{N}(t), m)N(t, m), \tag{4.75}$$

$$n(t, m, 0) = 2p(t, h(m), \tau)h'(m). \tag{4.76}$$

The system (4.73) and (4.74) provides conservation equations for $p(t,m,a)$ and $n(t,m,a)$ and can be derived in the same way as the continuity equation. Equations (4.75) and (4.76) describe the cellular flux between phases. Integrating both sides of (4.73) and (4.74) over the age variable a and using conditions (4.75) and (4.76) one can derive equations for the functions $N(t,m)$ and $P(t,m) = \int p(t,m,a)\,da$. To make the presentation more transparent, we assume additionally that δ, γ and β do not depend on m. Let $m(t)$ be the solution of the equation $m' = V(m)$ satisfying the initial condition $m(0) = m_0$ and let $g(m_0) = m(-\tau)$, $k(m) = g(h(m))$. Then

$$\begin{aligned}\frac{\partial N}{\partial t} + \frac{\partial (VN)}{\partial m} = &- (\delta + \beta(\bar{N}))N \\ &+ 2e^{-\gamma\tau}\beta(\bar{N}(t-\tau))k'(m)N(t-\tau, k(m)).\end{aligned} \tag{4.77}$$

Equation (4.77) is rather difficult to study because it is a partial differential equation in which there is an explicit temporal retardation as well as a nonlocal dependence in the maturation variable due to cell replication. Integrating both sides of (4.77) over m we obtain

$$\bar{N}'(t) = -(\delta + \beta(\bar{N}))\bar{N} + 2e^{-\gamma\tau}\beta(\bar{N}(t-\tau))\bar{N}(t-\tau). \tag{4.78}$$

In [75] it is proved the following result.

Theorem 4.11. *Assume that the delay Eq. (4.78) has a constant solution $\bar{N}_0 > 0$ and $\bar{N}_0$ is globally asymptotically stable. If*

$$(\delta + \beta(\bar{N}_0)) \log k'(0) < V'(0) \tag{4.79}$$

then there exists a stationary solution $N_0(m)$ of Eq. (4.77) and for every solution $N(t,m)$ of it we have

$$\lim_{t\to\infty} \int |N(t,m) - N_0(m)|\,dm = 0. \tag{4.80}$$

Condition (4.79) has an interesting biological interpretation. It shows that the stability of the population depends on the dynamics of low mature (small) cells. The term $k'(0)$ describes the relation between the maturation of the mother and daughter cells. If m is the maturation of a small mother cell at the moment of entering the proliferating phase, then the maturation of a new-born daughter cell is $m/k'(0)$. The term $c = \delta + \beta(N_0)$ is the rate of leaving of the resting phase (by being lost or by entering the proliferating phase). Since $V'(0)$ is the rate at which small cells mature, condition (4.79) means that the maturation of a big part of small cells will increase in the next generation.

4.7. Conclusion

In this lecture we presented some number of different types of structured models which we divided into four different classes according to the type of structure and time (discrete and continuous). The most advanced models are continuous time-structure models. They are usually described by one or by a system of partial differential equations (transport equations) with specific reproductions terms (non-local operators or integral boundary conditions). Advanced structured models contain also time delay (*e.g.* the delay connected with the cell-cycle, reproduction process, etc.), nonlinear terms connected with limited resources or second order terms connected with stochastic movement of individuals or stochastic noise if, for example, a parameter is some phenotype property and evolution is influenced by mutation.

We restrict our mathematical results to study simple asymptotic properties of models as asynchronous exponential growth and asymptotic stability. But our models can have more complicated behavior which can be studied using theoretical methods of dynamical systems. We can investigate such properties of models as the existence of a limit cycle, bifurcation, existence of invariant measures and chaos. It is hardly known that solutions of simple linear partial differential equations behave in a chaotic way. The following equation with the initial condition:

$$\frac{\partial u}{\partial t} + x\frac{\partial u}{\partial x} = \lambda u,$$
$$u(0,x) = v(x)$$

defines a dynamical system on the space $X = C[0,1]$ given by

$$S^t v(x) = u(t,x) = e^{\lambda t} v(e^{-t}x),$$

which is chaotic, practically in each sense of the meaning of this word. For example, for $\lambda > 0$ there exists a Gaussian measure with the support X invariant under $\{S^t\}_{t\geq 0}$ and the system is mixing [76] (see also a review [77]). This implies the topological chaos: the existence of dense trajectories (topological transitivity) and instability of trajectories. The topological chaos was also proved for the birth-death-type model [78].

In structured population models we mainly investigate the long time behavior of densities. As we show in these lectures there are a lot of results concerning asymptotic stability and asynchronous exponential growth, but there are no results concerning chaotic behavior on the set of densities. One of the reasons for the lack of such results is that there are no good mathematical tools to investigate this problem. Methods of studying chaos based on the paper [79] do not work in this case because they are strictly connected with dynamical systems acting on the whole linear spaces. Methods based on invariant measures seem to be too difficult in these models.

Another interesting problem is formation of spatial aggregates as a result of interaction between individuals. This phenomenon concerns a variety of biological species, from colonies of bacteria, swarms of larvae, or adult insects to fish schools, bird flocks etc. Numerous mathematical models have been proposed for these processes, *e.g.* [80, 81], but the problem of spatial heterogeneity is still far away from the real solution. The explanation, originally proposed by Turing [82] and developed in many papers, that diffusion can destabilise homogeneous distribution to produce pattern is often treated by biologists as a mathematical trick but not the real reason. Models based on interaction between individuals such as chemotaxis for simple organisms and social actions of animals should better explain the behavior of the whole population. Unfortunately it is not easy to build such models and very difficult to deduce from them qualitative and quantitative results which can be compared with biological observations.

There are also a lot of open problems in population genetics. Most of them are connected with stochastic models which are not discussed in these lectures, but also the models discussed in these lectures needs further investigations. For example, according to my knowledge, the Penna model (see Example 4.5) was not too intensively mathematically investigated. It would be interesting to show the existence and stability of a stationary distribution in this model and compare it with the age distribution in the human population. The models involving telomere lost (Examples 4.3 and 4.17) need future modifications which should include the recent knowledge concerning this subject. Also the model concerning the evolution of paralog families (Example 4.7) needs modifications and further studies. Genome evolution is a very complicated stochastic process which involves many events in addition to the ones considered in this model. We also ignore the possibility of dependence of rates of elementary events on the gene length, its location in the genome, genome size or the functional importance of a given gene.

Acknowledgements

This research was partially supported by the State Committee for Scientific Research (Poland) Grant No. N N201 0211 33 and by EC FP6 Marie Curie ToK programme SPADE2, MTKD-CT-2004-014508 and Polish MNiSW SPB-M.

References

[1] S. Anita, *Analysis and Control of Age-Dependent Population Dynamics*, Mathematical Modelling: Theory and Applications, vol. 11, Kluwer Academic Publishers, Dordrecht, (2000).
[2] B. Charlesworth, *Evolution in Age-Structured Populations.* (Cambridge Univ. Press, Cambridge, 1980).
[3] B. D. Chepko-Sade and Z. T. Halpin, Eds., *Mammalian Dispersal Patterns, the Effects of Social Structure on Population Genetics.* (University of Chicago Press, Chicago, 1987).
[4] J. M. Cushing, *An Introduction to Structured Population Dynamics*, Conference Series in Applied Mathematics, vol. 71, SIAM, Philadelphia, (1998).
[5] M. Iannelli, *Mathematical Theory of Age-Structured Population Dynamics.* (C.N.R. Giardini, Pisa, 1994).
[6] J.A.J. Metz and O. Diekmann, Eds., *The Dynamics of Physiologically Structured Populations,* (Springer Lecture Notes in Biomathematics, vol. 68, New York, 1986).
[7] S. Tuljapurkar and H. Caswell, Eds., *Structured-Population Models in Marine, Terrestrial, and Freshwater Systems.* (Chapman & Hall, New York, 1997).
[8] G. W. Webb, *Theory of Nonlinear Age-Dependent Population Dynamics.* (Marcel Dekker, New York, 1985).
[9] B. Frier, Roman life expectancy: Ulpian's evidence, *Harvard Studies in Classical Philology* **86**, 213–251, (1982).
[10] W. Mays, Ulpian's Table, *Actuarial Research Clearing House*, **2**, 95–122, (1979).
[11] E. Halley, An estimate of the degrees of mortality of mankind drawn from curious tables of births and funerals in the city of Breslau with an attempt to ascertain the price of annuities upon lives, *Phil. Trans.R. Soc. Lond.* **17**, 596–610, (1793).
[12] E. Halley, Some further considerations on the Breslau bills of mortality, *Phil. Trans. R. Soc. Lond.* **17**, 654–656, (1793).
[13] T. R. Malthus, *An Essay on the Principle of Population.* (London, 1798).
[14] N. Keyfitz and W. Flieger, *World Population Growth and Ageing: Demographic Trends in the Late Twentieth Century.* (University of Chicago Press, Chicago, 1990).
[15] B. Gompertz, On the nature of the function expressive of the law of human

mortality, and on a new mode of determining the value of life contingencies, *Phil. Trans. R. Soc. Lond.* **115**, 513–583, (1825).
[16] P. F. Verhulst, Notice sur la loi que la population suit dans son accroissement, *Corresp. Math. Phys.* **10**, 113–121, (1838).
[17] W.C. Allee, *Animal Aggregations.* (University of Chicago Press, Chicago, 1931).
[18] V. Volterra, Variazioni e fluttuazioni del numero d'individui in specie animali conviventi, *Mem. R. Accad. Naz. dei Lincei* **2**, 31–113, (1926).
[19] V. Volterra, Fluctuations in the abundance of a species considered mathematically, *Nature* **118**, 558–560, (1926).
[20] A. J. Lotka, *Elements of Physical Biology*, (Williams & Wilkins Co., Baltimore, 1925).
[21] A. N. Kolmogorov, Sulla teoria di Volterra della lotta per l'esistenza, *Giornale dell Istituto Italiano Degli Attuari* **7**, 74–80, (1936).
[22] F.M. Scudo and J.R. Ziegler, Eds., *The Golden Age of Theoretical Ecology: 1923 - 1940,* Lecture Notes in Biomathematics, vol. 22, Springer-Verlag, Berlin-Heidelberg-New York, (1978).
[23] V. Krivan, *Population dynamics: prey-predator models*, In eds. S.E. Jorgensen, *Encyclopedia of Ecology*, Elsevier, Oxford (2008 in press).
[24] W. O. Kermack and A. G. McKendrick, A contribution to the mathematical theory of epidemics, *Proc. Roy. Soc.* 115, 700–721, (1927).
[25] A.G. McKendrick, Application of mathematics to medical problems, *Proc. Edinb. Math. Soc.* 14, 98–130, (1926).
[26] F.R. Sharpe and A.J. Lotka, A problem in age-distributions, *Phil. Mag.* **21**, 435–438, (1911).
[27] J.D. Murray, *Mathematical Biology, Volume I: An Introduction and Volume II: Spatial Models and Biomedical Applications.* (Springer-Verlag, New York, 2002).
[28] H. R. Thieme, *Mathematics in Population Biology.* (Princeton University Press, Princeton and Oxford, 2003).
[29] J. Hofbauer and K. Sigmund, *The Theory of Evolution and Dynamical Systems. Mathematical aspects of selection.* (Cambridge Univ. Press, Cambridge, 1988).
[30] F. Brauer and C. Castillo-Chavez, *Mathematical Models in Population Biology and Epidemiology.* Texts in Appl. Math. 40, Springer, New York, (2001).
[31] M. Farkas, *Dynamical models in biology.* (Academic Press, Inc., San Diego 2001).
[32] J. F. Crow and M. Kimura. *An introduction to population genetics theory.* (Harper & Row, New York, 1970).
[33] W. J. Ewens, *Mathematical Population Genetics.* (Springer-Verlag, Berlin, 2004).
[34] R. A. Fisher, *The Genetical Theory of Natural Selection.* (Dover Publications, New York, 1958).
[35] D. Hartl and A. Clark, *Principles of Population Genetics.* (Sinauer Associates, Sunderland, 1997).
[36] M. Kimura, *The neutral theory of molecular evolution.* (Cambridge Univer-

sity Press, Cambridge, 1983).

[37] P. A. P. Moran, *The statistical processes of evolutionary theory.* (Clarendon Press, Oxford, 1962).

[38] T. Nagylaki, *Introduction to theoretical population genetics.* (Springer Verlag, Berlin, 1992).

[39] R. A. Blythe and A. J. McKane, Stochastic models of evolution in genetics, ecology and linguistics, *J. Stat. Mech.: Theor. Exp.* P07018, (2007). `http://www.iop.org/EJ/abstract/1742-5468/2007/07/P07018`.

[40] P. H. Leslie, On the Use of Matrices in Certain Population Mathematics, *Biometrika* **33**, 183–212, (1945).

[41] H. H. Schaefer, *Banach Lattices and Positive Operators.* (Springer-Verlag, Berlin, 1974).

[42] E. Goles, O. Schulz and M. Markus, Prime number selection of cycles in a predator-prey model, *Complexity* **6**, 33–38, (2001).

[43] T. J. P. Penna, A bit string model for biological aging, *J. Stat. Phys.* **78**, 1629–1633, (1995).

[44] R. M. C. de Almeida, S. Moss de Oliveira, and T. J. P. Penna, Theoretical approach to biological aging, *Physica A* **253**, 366–378, (1998).

[45] R. Rudnicki, J. Tiuryn and D. Wójtowicz, A model for the evolution of paralog families in genomes, *J. Math. Biology* **53**, 759–770, (2006).

[46] P.P. Slonimski, M.O. Mosse, P. Golik, A. Henaût, Y. Diaz, J.L. Risler, J.P. Comet, J.C. Aude, A. Wozniak, E. Glemet and J.J. Codani, The first laws of genomics, *Microbial Comp. Genomics* **3**, 46, (1998).

[47] M. A. Huynen and E. van Nimwegen, The frequency distribution of gene family size in complete genomes, *Molecular Biology Evolution* **15**, 583–589, (1998).

[48] J. Norris, *Markov Chains.* Cambridge Series on Statistical and Probabilistic Mathematics, Cambridge University Press, Cambridge, (1997).

[49] A. Bobrowski, *Functional analysis for probability and stochastic processes. An introduction.* (Cambridge University Press, Cambridge, 2005).

[50] E. Hille, R. S. Phillips, *Functional analysis and semi-groups.* American Mathematical Society Colloquium Publications **31**, American Mathematical Society, Providence, R. I., (1957).

[51] A. Lasota, J. A. Yorke, Exact dynamical systems and the Frobenius-Perron operator, *Trans. AMS* **273**, 375–384, (1982).

[52] K. Pichór and R. Rudnicki, Continuous Markov semigroups and stability of transport equations, *J. Math. Anal. Appl.* **249**, 668–685, (2000).

[53] R. Rudnicki, On asymptotic stability and sweeping for Markov operators, *Bull. Pol. Ac.: Math.*, **43**, 245–262, (1995).

[54] A. Lasota and M. C. Mackey, Globally asymptotic properties of proliferating cell populations, *J. Math. Biol.* **19**, 43–62, (1984).

[55] J.J. Tyson and K.B. Hannsgen, Cell growth and division: A deterministic/probabilistic model of the cell-cycle, *J. Math. Biol.* **23**, 231–246, (1986).

[56] J. Tyrcha, Asymptotic stability in a generalized probabilistic/deterministic model of the cell-cycle, *J. Math. Biology* **26**, 465–475, (1988).

[57] H. Gacki and A. Lasota, Markov operators defined by Volterra type integrals

with advanced argument, *Ann. Polon. Math.* **51**, 155–166, (1990).
[58] K. Łoskot and R. Rudnicki, Sweeping of some integral operators, *Bull. Pol. Ac.: Math.* **37**, 229–235, (1989).
[59] R. Rudnicki, Stability in L^1 of some integral operators, *Integral Equation Operator Theory* **24**, 320–327, (1996).
[60] H. Inaba, Weak ergodicity of population evolution processes, *Math. Biosci.* **96**, 195–219, (1989).
[61] H. Inaba, Strong ergodicity for perturbed dual semigroups and application to age-dependent population dynamics, *J. Math. Anal. Appl.* **165**, 102–132, (1992).
[62] R. Rudnicki and M. C. Mackey, Asymptotic similarity and Malthusian growth in autonomous and nonautonomous populations, *J. Math. Anal. Appl.* **187**, 548–566, (1994).
[63] M. Ważewska–Czyżewska and A. Lasota, Matematyczne problemy dynamiki układu krwinek czerwonych, *Roczniki PTM, Matematyka Stosowana* **6**, 23–40, (1976).
[64] L. Glass and M. C. Mackey, *From Clocks to Chaos, The Rhythms of Life.* (Princeton University Press, Princeton, 1988).
[65] G. I. Bell and E. C. Anderson, Cell growth and division I. A Mathematical model with applications to cell volume distributions in mammalian suspension cultures, *Biophysical Journal* **7**, 329–351, (1967).
[66] O. Diekmann, H. J. A. M. Heijmans and H. R. Thieme, On the stability of the cell size distribution, *J. Math. Biology* **19**, 227–248, (1984).
[67] M. Gyllenberg and H. J. A. M. Heijmans, An abstract delay-differential equation modelling size dependent cell growth and division, *SIAM J. Math. Anal.* **18**, 74–88, (1987).
[68] R. Rudnicki and K. Pichór, Markov semigroups and stability of the cell maturation distribution, *J. Biol. Systems* **8**, 69–94, (2000).
[69] G. W. Webb, Structured population dynamics, In eds. R. Rudnicki, *Mathematical Modelling of Population Dynamics,* Banach Center Publication **63**, pp. 123–163, Warszawa (2004).
[70] S.I. Rubinow, A maturity time representation for cell populations, *Biophy. J.* **8**, 1055–1073, (1968).
[71] R. M. C. de Almeida and G. L. Thomas, Scaling in a continuous time model for biological aging, *Int. J. Mod. Phys. C.* **11**, 1209–1224, (2000).
[72] O. Arino, E. Sánchez, and G. F. Webb, Polynomial growth dynamics of telomere loss in a heterogeneous cell population, *Dynam. Contin. Discrete Impuls. Systems* **3**, 263–282, (1997).
[73] T. Lipniacki, P. Paszek, A. Marciniak-Czochra, A.R. Brasier, M. Kimmel, Transcriptional stochasticity in gene expression, *J. Theor. Biol.* **238**, 348–367, (2006).
[74] A. Bobrowski, T. Lipniacki, K. Pichór, and R. Rudnicki, *Asymptotic behavior of distributions of mRNA and protein levels in a model of stochastic gene expression*, *J. Math. Anal. Appl.* **333**, 753–769, (2007).
[75] M.C. Mackey and R. Rudnicki, Global stability in a delayed partial differential equation describing cellular replication, *J. Math. Biol.* **33**, 89–109,

(1994).
[76] R. Rudnicki, Strong ergodic properties of a first-order partial differential equation, *J. Math. Anal. Appl.* **133**, 14–26, (1988).
[77] R. Rudnicki, Chaos for some infinite-dimensional dynamical systems, *Math. Meth. Appl. Sci.* **27**, 723–738, (2004).
[78] J. Banasiak and M. Lachowicz, Topological chaos for birth-and-death-type models with proliferation, *Math. Models Methods Appl. Sci.* **12**, 755–775, (2002).
[79] W. Desch, W. Schappacher and G. F. Webb, Hypercyclic and chaotic semigroups of linear operators, *Ergodic Theory Dynamical Systems* **17**, 793–819, (1997).
[80] D. Morale, V. Capasso and K. Oelschläger, An interacting particle system modelling aggregation behavior: from individuals to populations, *J. Math. Biol.* **50**, 49–66, (2005).
[81] A. Okubo and S. Levin, Eds., *Diffusion and Ecological Problems: Modern Perspectives.* (Springer Verlag, New York, 2001).
[82] A. M. Turing, On the chemical basis of morphogenesis, *Philos. Trans. Roy. Soc. London B* **237**, 37–72, (1952).

Chapter 5

Age-structured population models with genetics

Mirosław R. Dudek[1] and Tadeusz Nadzieja[2]

[1] *Institute of Physics, University of Zielona Góra,*
ul. Z. Szafrana 4a,
PL-65516 Zielona Góra, Poland,
mdudek@proton.if.uz.zgora.pl

[2] *Faculty of Mathematics, Computer Science and Econometrics,*
University of Zielona Góra,
ul. Z. Szafrana 4a,
PL-65516 Zielona Góra, Poland,
T.Nadzieja@wmie.uz.zgora.pl

Some mathematical models of time evolution of the age-structured biological populations have been discussed. It has been shown that the genetic information represented by the distribution of the defective genes in the population under consideration can be included into the deterministic models which use ordinary differential equations or delay differential equations. The results for the equilibrium genetic populations have been compared with the equilibrium populations in the Penna model of genetic evolution. It has been stressed the difference between the models which use ordinary differential equations and delay differential equations.

Contents

5.1. Introduction

Mathematical modeling the evolution of biological population has a long history (see Chap. 4). The mathematical tools used in the investigation of biological processes are very rich: deterministic methods (ordinary, partial, delay differential equations, integral equations), stochastic methods, dynamical systems, discrete mathematics and numerical analysis. The aim of our lecture is to describe some models of time evolution of the biological population. We take into consideration the age structure of the population under consideration and relate it to genetic information. In order to model the evolution of such population we modify the classical growth models. We introduce into them the age structure and/or the delay. The feature of the solutions obtained from the delay differential equations (DDE) are essentially different from the solutions of the ordinary differential equations (ODE). Even in the simplest case of the Malthus-Verhulst equations if we introduce the delay (time lag) into the Verhulst factor we can obtain the periodic solution, which does not exist in the case of the Malthus-Verhulst equation. Introducing the delay into the few-variables models leads to the new mathematical phenomena, which are observed in real biological precesses. In Sec. 5.4 we show that tedious and long calculation in classical Penna model can be replaced by a relatively simple calculations of the solutions of a system of ordinary differential equations. In the last section we propose the new numerical method for solving the ODE and DDE appearing in mathematical modeling.

5.2. One-variable models, single species

An extensive literature deals with modeling biological populations with ordinary and partial differential equations (ODE and PDE). For the particular case of the well-known Lotka-Volterra model [1, 2] one gets over a few dozen thousand hits on the internet for the search pattern: *Lotka-Volterra "differential equation"*. The population differential equations can describe various stages of the populations under consideration, even the phase transitions. In this section we start from one-dimensional (one-variable) equation relating to a single species.

The simplest example of the population model was introduced by Thomas R. Malthus in XVIII century for the population density $N(t)$, which describes the population growth proportional to the current population. In terms of differential equations it is expressed in the following way:

$$\frac{dN(t)}{dt} = rN(t), \quad N(0) = N_0. \tag{5.1}$$

The symbol r denotes the effective population growth rate per unit time and N_0 is the initial value of the population density. This rate is the difference of the individual's birth rate and death rate. The solution of Eq. (5.1) shows an exponential growth of the population for $t \geq 0$

$$N(t) = N_0 e^{rt}. \tag{5.2}$$

The phrase "Malthusian growth" which is often met in literature is a synonym of this exponential growth. However, in a finite ecosystem the species cannot grow infinitely and it is expected a kind of saturation for the value of $N(t)$. Usually, in order to prevent the exponential growth of $N(t)$ the value of $N(t)$ is multiplied by the Verhulst factor $(1 - N(t)/\Omega)$ where Ω represents the environmental capacity for the population under consideration. The corrected equation takes the following form:

$$\frac{dN(t)}{dt} = rN(t)\left(1 - \frac{N(t)}{\Omega}\right), \tag{5.3}$$

which is known as the logistic differential equation. Another form of this equation

$$\frac{dN(t)}{dt} = rN(t) - mN^2(t) \tag{5.4}$$

is called the Malthus-Verhulst equation. In this case, $\Omega = r/m$ and the symbol m denotes the death rate due to interactions with other individuals in the population under consideration. For $t \geq 0$, the solution of Eq. (5.4) with the initial condition $N(0) = N_0$ is [2]

$$N(t) = \frac{N_0 e^{rt}}{1 + N_0 \frac{m}{r}(e^{rt} - 1)}, \tag{5.5}$$

where two asymptotic scenarios are possible for $r < 0$ and $r > 0$ as $t \to \infty$. In the first case the population becomes extinct, $N(t) \to 0$, whereas in the latter case the population density $N(t)$ tends to the value r/m.

In some applications, *e.g.*, during modelling the growth of the number of tumor cells, it is much better to use the Gompertz factor $-\ln(\frac{N_1(t)}{\Omega})$

instead of the Verhulst factor. These two factors can be related by the following Taylor series expansion (page 68 in [3]):

$$-\ln\left(\frac{N(t)}{\Omega}\right) = \left(1 - \frac{N(t)}{\Omega}\right) + \frac{1}{2}\left(1 - \frac{N(t)}{\Omega}\right)^2 + \ldots, \tag{5.6}$$

which is valid for $|\frac{N(t)}{\Omega}| < 1$. Notice, that if the series was used in Eq. (5.3) instead of the Verhulst multiplier then the higher order intra-species specific interactions would be present.

The simplified one-variable equations like Eq. (5.3) can easily be extended to the case when they are history dependent, *e.g.* through the delay differential equations (DDE). Hutchinson [4], was one of the first to introduce DDE to population dynamics. It was a simple extension of the logistic differential equation (Eq. (5.3))

$$\frac{dN(t)}{dt} = rN(t)\left(1 - \frac{N(t-\tau)}{\Omega}\right), \quad N(t) = N_0(t) \text{ for } t \in [-\tau, 0], \tag{5.7}$$

by some time lag τ (time delay) and $N_0(t)$ is a given continuous function on time interval $[-\tau, 0]$. The purpose of introducing the delay time was to include, in some way, the hatching and maturation periods into the population growth dynamics. Hutchinson suggested that the finite delay time is responsible for the observed oscillations in many biological systems. A few examples of the phenomenon can be the Cheyne-Stokes breathing - oscillations of breathing frequency related to carbon dioxide level in the blood observed in cardiac diseases (*e.g.* [5] – [8]), time delay in the production of white blood cells [9], the dynamics of the system of red blood cells [10], and time delay between variations in heart rate [11]. Common examples of time delay in the prey-predator system is gestation time for predator after consuming prey or the effect of delayed environmental interactions on the species dynamics which is introduced in the same manner as the Verhulst factor in Eq. (5.7). A short review on the applications of DDE in biosciences can be found in [12]. It is often the case that the time delay of the system under consideration is unknown and then one has to guess the value of τ from time series data. An example of the graphical method for estimating time delay in the relationship between the population and environment can be found in [13]. Another possible approach is some minimization procedure for the set of unknown parameters for a general form of the delay model appropriate for the populations under consideration [8].

In the case of the logistic differential equation with no delay ($\tau = 0$) the analytical solution of Eq. (5.7) is given by Eq. (5.5). It is monotonic and

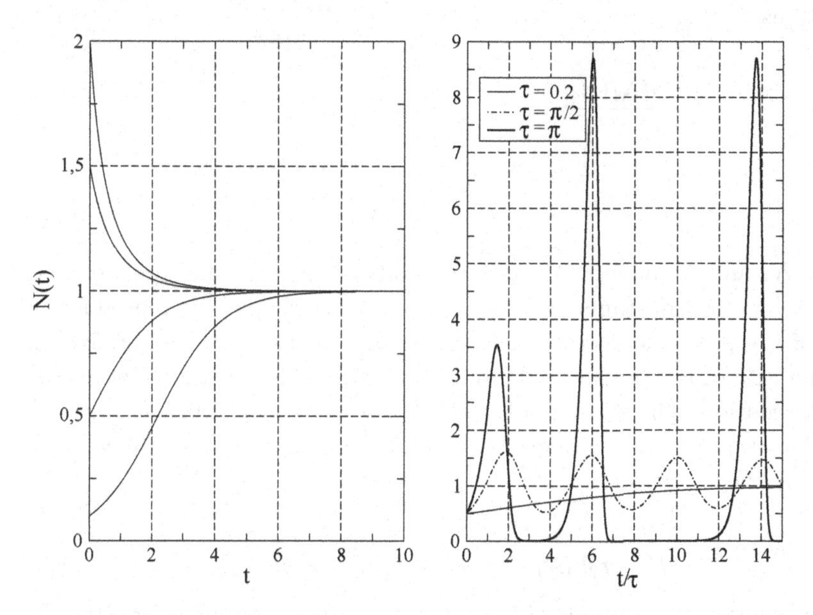

Fig. 5.1. Left panel: solutions of the logistic differential equation, Eq. (5.7) for $\tau = 0$, for different initial condition $N_0 = 0.1, 0.2, 0.5, 1.5, 2.0$. Right panel: solutions of Eq. (5.7) for $\tau = 0.2, \pi/2, \pi$ for initial condition $N_0(t) \equiv 0.5$ in $[-\tau, 0]$. In both panels $r = 1$, $\Omega = 1$. The solutions have been obtained with the help of the numerical method introduced in Sec. 5.5.

converging to the value of Ω. A typical example of such solutions are the plots in the left panel of Fig. 5.1. These plots have been presented for $\Omega = 1$ but the value of Ω can be any positive finite number appropriate to the size of the ecosystem under consideration. Unlike the monotonic solutions for $\tau = 0$ the solutions for $\tau > 0$ can start to oscillate. The solution of Eq. (5.7) monotonically converges to the value of Ω for $0 < r\tau < \frac{1}{e}$, it converges in an oscillatory fashion to the value of Ω for $\frac{1}{e} < r\tau < \frac{\pi}{2}$ and it oscillates in a stable limit cycle for $r\tau > \frac{\pi}{2}$ ([14], [15] and [8]). Some examples of the different types of solutions can be observed in the right panel of Fig. 5.1.

5.3. A few-variable model

The population models are usually represented by various types of the predator-prey systems. The classic example is the phenomenological Lotka-Volterra model which describes the interaction of two species, a prey and a predator, and this interaction is modeled with the following differential equations:

$$\frac{dN_1(t)}{dt} = r_1N_1(t) - m_1N_1(t)N_2(t),$$
$$\frac{dN_2(t)}{dt} = -r_2N_2(t) + m_2N_1(t)N_2(t), \tag{5.8}$$

where N_1 and N_2 denote population density of the prey and predator, r_1, m_1, r_2, m_2 are constants and they denote, respectively, the reproduction rate of the prey, the decrease rate of the prey due to the prey predation, decline rate of the predator, reproduction rate of the predator due to the prey predation. The system of Eqs. (5.8) has a singular point $N_1 = r_2/m_2$ and $N_2 = r_1/m_1$ and there exists the first integral of motion [16] which is given by the expression:

$$H = r_2 \mathrm{ln} N_1 - m_2N_1 + r_1 \mathrm{lnN}_2 - m_1N_2. \tag{5.9}$$

Consequently, the solutions of Eqs. (5.8) are either periodic or constant.

In a finite ecosystem the prey species cannot grow infinitely as is suggested by Eqs. (5.8) in the absence of predator ($N_2 = 0$). The saturation mechanism can be introduced with the help of the Verhulst factor and the new extended form of the Lotka-Volterra equations

$$\frac{dN_1(t)}{dt} = r_1N_1(t)\left(1 - \frac{N_1(t)}{\Omega}\right) - m_1N_1(t)N_2(t),$$
$$\frac{dN_2(t)}{dt} = -r_2N_2(t) + m_2N_1(t)N_2(t), \tag{5.10}$$

is substituted for the Eqs. (5.8). There is no first integral of motion in Eqs. (5.10) (Gazis *et al.* [16]).

The choice of the saturation factor also depends on the type of the species under consideration *i.e.* whether it is prey or predator. In particular, the growth controlling the factor for N_2 in Eqs. (5.10) should include the predator's specifity like the limited consuming capacity of prey per day, *i.e.*, it at least should be proportional to N_1 for small values of N_1 and it should be constant for large N_1. Some authors use the Watt multiplier [16, 17] which is given by the expression $\frac{m_1}{c}\left(1 - e^{-cN_1(t)}\right)$ where c is a

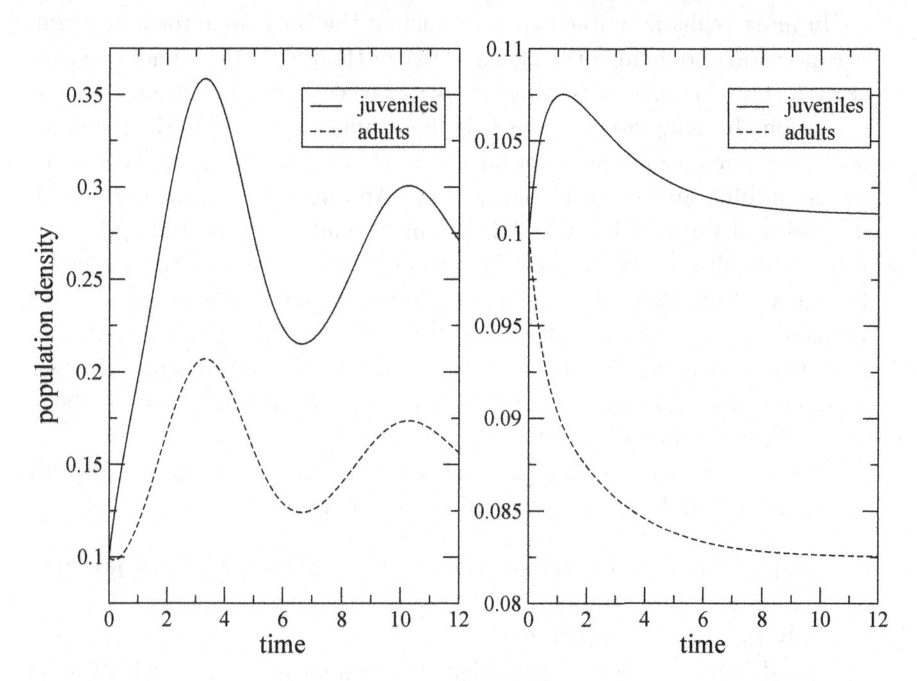

Fig. 5.2. Dependence of the density N_1 (juveniles) and N_2 (adults) in Eqs. (5.12) on time in the case of the time delay $\tau = 3$ (left) and $\tau = 1.5$ (right). In both panels $\Omega = 1$. The solutions N_1 and N_2 have been obtained with the help of the numerical method introduced in Sec. 5.5.

constant which scales the predation capacity. Then, the species interaction term $N_1(t)N_2(t)$ in both differential equations in Eqs. (5.10) takes a new form:

$$N_1(t)N_2(t) \rightarrow \frac{N_2(t)}{c}\left(1 - e^{-cN_1(t)}\right). \tag{5.11}$$

The set of differential equations (5.8) or (5.10) can be easily generalized to the multispecies case. Then, the chaotic solutions are possible (*e.g.* [18] – [20]). The existence of chaos became evident since the work of May [21, 22]. Typical example of the mulispecies analyses takes place when the species are related according to food-chain. In the case of the nearest-neighbor relation the species $i+1$ preys on species i. More complicated prey-predator relations were simulated with a 5×5 Chowdhury lattice [23] with a fixed number of six food levels. The application of the multispecies relations to fisheries management can be found in paper by Harvey *et al.* [24].

In more realistic mathematical modeling the prey-predator interaction it is necessary to include the age-specifity of the populations under consideration. For example, various age groups of prey show different reactions to predation. In many applications it is then sufficient to divide the individuals in the population under consideration into a few age-groups, *e.g.*, very young, adults, and the very old or sick. Another observation is that the dynamics of the growing population can depend on the initial age structure, especially in the case of the predator-prey relation (*e.g.*, paper by Hance and Impe [25]). Some examples of the papers on the age-structured Lotka-Volterra system can be found in [6, 16, 26–28]. These models can be enriched by some time lag τ with respect to the environmental changes or inter-species relations (*e.g.*, [29]), in a way similar to what has been discussed in the previous section.

Let us consider a simple generalization of the single species growth model in Eq. (5.7) to the age-specific case where:

- the population under consideration consists of two age groups: juveniles (N_1) and adults (N_2),
- only the adults can reproduce,
- the density of the newborn offspring is equal to $\alpha_0 N_2(t)$ where α_0 is the population reproduction coefficient,
- the graduation of the individuals from the age group $a-1$ to the age group a is only the ageing process and individuals older than age $a=2$ are die,
- the growth in number of the whole population (juveniles and adults) is controlled by the Verhulst factor $V(t) = 1 - \frac{N_1(t-\tau)+N_2(t-\tau)}{\Omega}$ with some delay time τ to account for the timing of events concerning reaction to an environment and intra-species relations.

The corresponding differential equations take the following form:

$$\frac{dN_1(t)}{dt} = \alpha_0 V(t-\tau) N_2(t) - N_1(t),$$
$$\frac{dN_2(t)}{dt} = V(t-\tau) N_1(t) - N_2(t), \tag{5.12}$$

where the change in time of the density $N_1(t)$ of juveniles consists of two terms representing the graduation of the newborn individuals to the juveniles and the graduation of the juveniles to adults, and the change in time of the density $N_2(t)$ of adults consists of the graduation of the juveniles to adults and the mortality term. Figure 5.2 shows how the densities of juveniles and adults depend on time in the case of different values of delay time. Notice that the various age groups of the individuals can respond differently to the applied delayed Verhulst factor although in this version of the model it is the same for juveniles and adults.

Other extentions can be helpful in modeling realistic populations including some specific details characterizing the population like parental care, inter-species interactions etc.

5.4. Models with genetics

The discussion of a predator-prey model with genetics was started by Ray *et al.* [30], who showed that the system passes from the oscillatory solution of the Lotka-Volterra equations into a steady-state regime, which exhibits some features of self-organized criticality. They used the Monte-Carlo computer simulations to mimic the Lotka and Volterra differential equations in discrete time steps. This method allows converting a phenomenological ODE system into the corresponding microscopic model. In this case, $N_1(t)$ and $N_2(t)$ are interpreted as integers representing number of individuals of the prey and predators. A computer experiment representing Eqs. (5.8) consists of the following steps:

(i) t=0; setting the input parameters: $N_1 = N_1(0)$, $N_2 = N_2(0)$, r_1, m_2, r_2, m_1, number of time steps T,
(ii) $t = t+1$; upgrading the numbers N_1 and N_2 with the help of the Monte-Carlo method where the the growth/death probability represents the respective rate coefficients r_1, m_2, r_2, m_1.
(iii) goto (ii) unless number of time steps is smaller than T.

The main difference between the results for $N_1(t)$ and $N_2(t)$ in computer simulations and phenomenological solutions of Eqs. (5.8) is that there appear fluctuations $\Delta N_1(t)$ and $\Delta N_2(t)$ in the number of individuals around their mean values $\langle N_1 \rangle$ and $\langle N_2 \rangle$, where $N_1(t) = \langle N_1 \rangle + \Delta N_1(t)$ and $N_1(t) = \langle N_2 \rangle + \Delta N_2(t)$. The computer experiment resembles a "real world" where population number fluctuations can even make the species extinct.

In the paper by Ray *et al.* [30], the genes were incorporated into the populations which were represented by a unit vector determining the fitness of each individual. When the vectors of two randomly chosen individuals from the prey and predator's population were diametrically opposed, then the prey had not been preyed on. Otherwise, the prey was preyed on and the highest probability of predation was when the vectors were colinear. In the model, the offspring has one parent (haploid population) and it inherits a copy of its parents genes.

A good example of the population model which describes genetic evolution was introduced by Penna [31, 32]. The Penna model (see Chap. 2) turned out to be very successful in interpreting the demographers data of real populations even as complex as human populations [32] – [35]. In the original asexual version of the Penna model [31], the population under consideration consists of individuals represented by genomes defined as a string of n bits. The bits represent states of genes where 0 denotes its functional allele and 1 its bad allele. There are many models in literature in which the bit-strings represent genome, *e.g.* [36] – [41]. However, the specifity of the Penna model of genetic evolution is that in this model all genes are switched on chronologically - each bit corresponds to one "year" - and there is a maximum life span of an individual which is equal to a_D "years". It is assumed that if an individual experiences T genetic diseases, *i.e.*, T bad alleles have been switched on until its present age, then it dies. After reaching the fertility age a_F an individual gives birth to some number offspring whose genomes are mutated versions of the parental genome. The mutation rate is constant. After the mutation the affected gene is represented by a bit with a value opposite to the value before the mutation.

Figure 5.3 shows the results of the Monte-Carlo simulations of a Penna population in the case when back mutations from 1 to zero are not allowed. The individuals in the population belong to different age groups and the three figures on the left panels show how many of defective genes specific for age a per individual are present in the corresponding age group. The figures in the right panels show what the fraction of age group specific for age a in the whole population is. The computer simulations started with 10% of bad genes in the population which have been equally distributed among different age groups. Additionally, all age groups were initially represented by the same number of individuals. This initial setup can be observed in two top figures. After long time of evolution the distribution of defective genes ε_a approaches the equilibrium form - it is well approximated by the shape in the bottom figure in the left panel of Fig. 5.3. Such a population is called equilibrium population. It can be seen from the right panel of Fig. 5.3 that the maximum age of individuals in this equilibrium population is equal

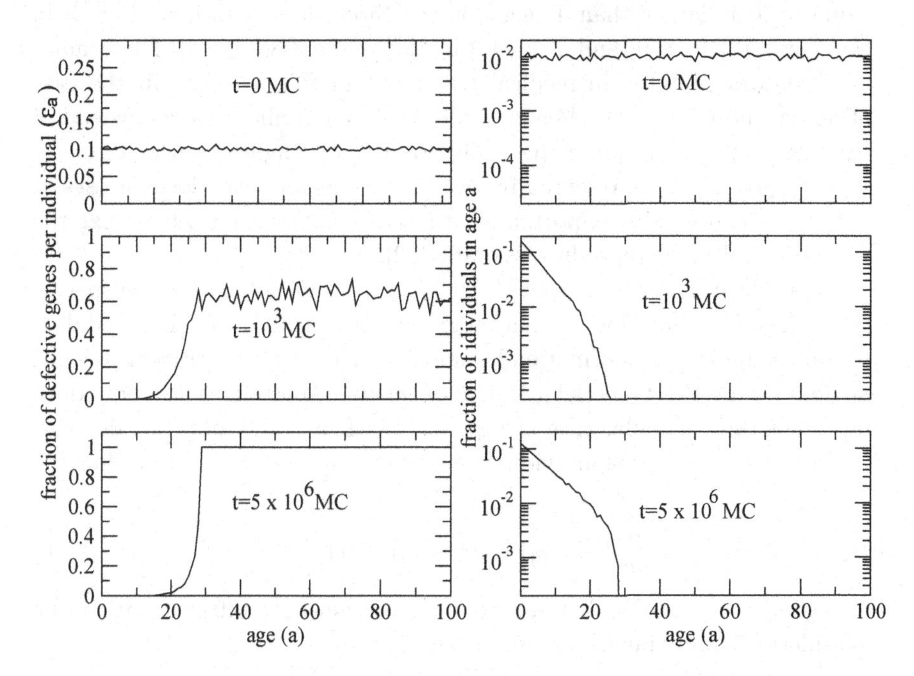

Fig. 5.3. Left: fraction ε_a of the defective genes specific for age a in the population per individual versus age a, Right: semilog plot of the fraction of individuals at age a in the whole population. The initial parameters: $N(0) = 10000$, $\Omega = 2 \times 10^5$, $a_F = 15$, length of the bit-string n=100, T=1.

to $a_D = 29$. It is the age for which the fraction of defective genes is equal to 1. In the case when the threshold for bad genes is set to the value $T = 1$ the fraction ε_a of defective genes specific for individual's age a in the equilibrium population consists of two parts:

$$\varepsilon_a = \frac{a^d e^{b(1+a_F-a_D)}}{1+a_F} \tag{5.13}$$

for $a \leq a_F$ and

$$\varepsilon_a = e^{b(a-a_D)} \tag{5.14}$$

for $a > a_F$ with some constants b and d. The detailed discussion of the expressions in Eq. (5.13) and Eq. (5.14) can be found in [42].

It was suggested in [34, 35] that in realistic populations the effective value of T is larger than 1, e.g, for the Swedish population: $T = 3$ in the years 1870–1879, and $T = 4.9$ in the years 1980–1989. These numbers confirm progress in medical treatment in this society. In the case of human population the Penna model has to describe diploids instead of haploids. The extension of the Penna model is straightforward - the individual's genome is represented by two bit-strings and each locus possesses two alleles. It is also important in the mathematical modeling that the human populations reproduce sexually [32].

In a recent paper by Dudek [42] it has been shown that the age-specific differential equations for population growth appear to be a good candidate to represent genetic information because the age structure introduces some analogy to the Penna model [31], [32] of genetic evolution. These equations represent the particular case of Eqs. (5.12) when $\tau = 0$ and they describe the change in number of individuals specific for age $a = 1, \ldots, a_D$ as follows:

$$\frac{dN_1(t)}{dt} = \alpha_0 V(t)(1 - \varepsilon_0)N_0(t) - N_1(t), \tag{5.15}$$

for age $a = 1$ where N_0 is the number of the newborn offspring (it can be considered as individuals specific for age $a = 0$)

$$N_0 = \sum_{a \geq aF} N_a(t) \tag{5.16}$$

and

$$\frac{dN_a(t)}{dt} = V(t)(1 - \varepsilon_{a-1})N_{a-1}(t) - N_a(t) \tag{5.17}$$

for age $a = 2, \ldots, a_D$, and V represents the Verhulst saturation factor

$$V(t) = 1 - \frac{1}{\Omega}\sum_{a=1}^{a_D} N_a(t). \tag{5.18}$$

with Ω being the saturation level. In the model, only adults can reproduce and the coefficient α_0 represents reproduction potential of the population under consideration. The first term in Eqs. (5.15) and (5.17) is the graduation term from the age $a - 1$ to the age a whereas the other part of these equations describes the graduation to the age $a + 1$ and in particular the individual's mortality if $a = a_D$. The Verhulst factor (the same as in the Penna model) ensures that the solution of Eq. (5.15) and Eq. (5.17) saturates at long times.

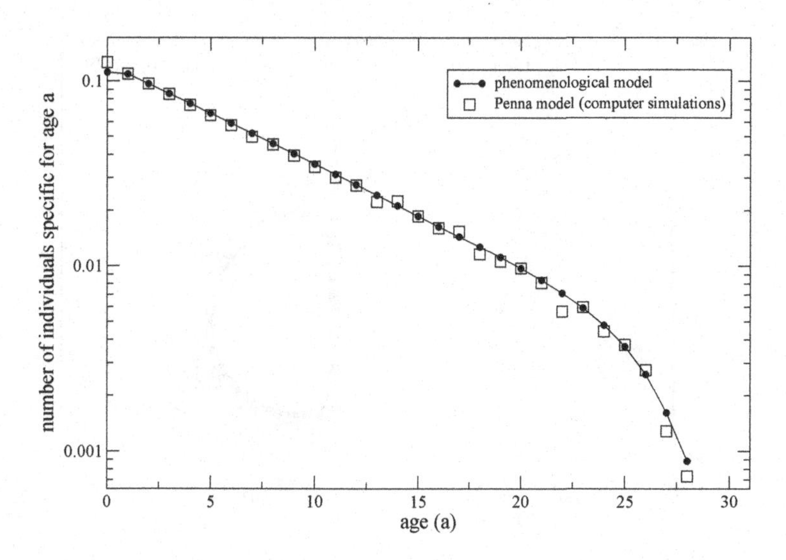

Fig. 5.4. Comparison of the computer simulations results for the Penna model and the corresponding results for the deterministic growth model in Eqs. (5.15) and (5.17). The plot of the fraction of individuals at age a in the whole population of individuals has been presented. The parameters for the Penna model are the same as in Fig. (5.3) whereas the parameters for the phenomenological model $a_F = 15$, $a_D = 29$, $d = 0.69$ and $b = 0.379$.

The asymptotic dependence of $N_a(t)$ on the age a obtained from Eq. (5.15) and Eq. (5.17) coincides well with the one which can be obtained from the corresponding equilibrium Penna population [42] for the same initial conditions and the reproduction coefficient α_0. This is a very important result because it means that instead of time consuming simulations of some population it is possible to guess its dynamical behavior with the help of the phenomenological equations in Eq. (5.15) and Eq. (5.17) unless the genetic information about the frequency of genetic diseases is known in the population. In this case is possible to determine the empirical values of the parameters b and d. An example of the equivalence of the results of the computer simulations and deterministic model can be observed in Fig. 5.4 where both the solution of the deterministic model Eq. (5.15) and Eq. (5.17) are plotted in a semi-log scale.

A series of examples of dynamical behavior of genetic populations which are defined similarly to in Eqs. (5.15) and (5.17) has been discussed in [42]. In order to show how important the genetic information can appear for the

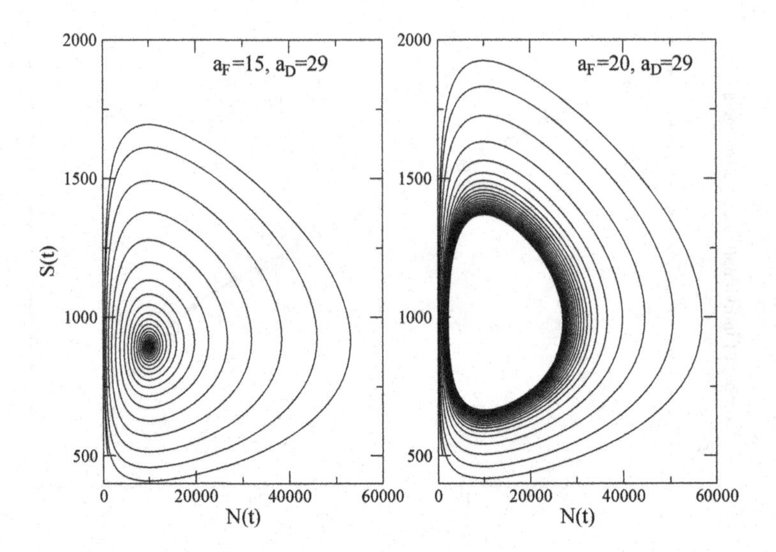

Fig. 5.5. Both panels show the variation of the number of individuals in the ecosystem consisting of the non age-structured prey $S(t)$ and age-structured predator. The difference in the panels is that on the left the predator's population has the fertility age $a_F = 15$ whereas on the right $a_F = 20$. The remaining parameters are the same, $a_D = 29$ and $\alpha_0 = 1.1$.

population dynamics it is tested in a simple prey-predator model:

$$\begin{aligned}
\frac{dS(t)}{dt} &= \alpha_S S(t)\,(1 - \frac{S(t)}{\Omega_S}) - m_S N(t) S(t), \\
\frac{dN_1(t)}{dt} &= \alpha_0 S(t)(1 - \varepsilon_0) N_0(t) - N_1(t), \\
\frac{dN_a(t)}{dt} &= S(t)(1 - \varepsilon_{a-1}) N_{a-1}(t) - N_a(t),
\end{aligned} \tag{5.19}$$

where $a = 2, \ldots, a_D$, the second equation corresponds to the case of age $a = 1$, N_0 is the same as in Eq. (5.16), $N(t)$ is the total number of predators

$$N(t) = \sum_{a=1}^{a_D} N_a(t), \tag{5.20}$$

$S(t)$ is the prey, Ω_S represents the saturation level for the prey population, the coefficient α_S is the regeneration rate of the prey, m_S is the damping coefficient due to predation. In order to follow the Penna model of genetic evolution the species evolves according to the Eqs. (5.15) and (5.17) but the predators' growth is controlled by the amount $S(t)$ of the available prey instead of the Verhulst factor, $V(t)$. However, a Verhulst term is assumed for the self-regeneration of the prey.

How complex the behavior of the model can be is shown in Fig. 5.5 where two qualitatively different asymptotic dynamics of the model are shown for the same initial conditions but different values of a_F: the spiraling to a fixed point (left panel) and converging to a limiting cycle (right panel). For example, imagine that according to some social regulations since a specific time the individuals in some equilibrium population start to reproduce at much older ages then a_F. The consequences of this shift in the reproduction age is that the solutions of Eqs. (5.19) can change qualitatively from the trend approaching the fixed point (equilibrium population) as in the left panel of Fig. 5.5 to oscillations (non-equilibrium population) as in the right panel of Fig. 5.5. It could happen that such a transition between these two types of solutions of the Lotka-Volterra equations Eqs. (5.19) could cause the species to become extinct.

It has been found [43] that Tasmanian devils (*Sarcophilus laniarius*; *Dasyuridae*), a population restricted to the island of Tasmania, have low genetic diversity and this can be related with large oscillations of population density. Recently, an increased intensity of infectious facial cancer in the population has been observed and this threatens Tasmanian devils with extinction [44]. The computer simulations performed in [45] suggested that population size oscillations can even change the shape of the function representing frequencies $\{\varepsilon_a\}_0^{a_D}$ of the defective genes in the population. In particular, in this paper the influence of the catastrophic reduction on the size of the genetic population was investigated. If the catastrophe was applied for many population generations then the distribution of the defective genes $\{\varepsilon_a\}_0^{a_D}$ lost its stability and typically the population become extinct.

5.5. Solving ODE and DDE in population model

There are important differences between the Cauchy problem for ODE

$$x'(t) = f(x(t), t), \quad x(t_0) = x_0, \quad x \in \Omega \subset R^n \tag{5.21}$$

and initial problem for DDE

$$x'(t) = f(x(t), x(t-\tau)), \quad x(t) = x_0(t), \quad \text{for } t \in [-\tau, 0]. \tag{5.22}$$

In the first one the space of initial conditions (a subset Ω of n- dimensional space) has finite dimension, for the second one the natural space of initial conditions is the set of continuous functions defined on interval, $[-\tau, 0]$ so the dimensionality of the space is infinite. Moreover, the solutions of Eq. (5.21) are smooth in contrast with the solution of Eq. (5.22), which are usually continuous only. To solve numerically the problem in Eq. (5.21) we have to do a discretization of the differential equation, in the problem in Eq. (5.22) we should do a discretization not only of the differential equation but also the initial condition.

Usually, differential or delay equations appearing in the models describing time evolution of biological populations are not integrable, *i.e.* the solutions can not be written in an explicit form. To obtain the information about the behavior of solutions we use numerical methods. Most of algorithms for ODE have finite difference form, like various Runge-Kutta algorithms [46]. The problem of solving DDE can be, in some sense, reduced to a problem for ODE. To solve the problem on the interval $[0, \tau]$ we consider the Cauchy problem of the form of Eq. (5.21)

$$x'(t) = F(x(t), t) := f(x(t), x_0(t-\tau)), \quad x(0) = x_0, \tag{5.23}$$

for which we can use the standard numerical methods for ODE.

The addition of delay time into the continuous population models makes the problem of the accuracy of numerical solution of DDE more complicated. We suggest another algorithm which is not a finite-difference method and its main advantage is in some cases even a few orders of magnitude smaller amount of rounding errors. There is no formal proof of the correctness of this algorithm, but it seems that our method gives satisfactory results for wide classes of problems of Eq. (5.21). Moreover, in some cases it can be much faster then the traditional numerical approach (for example for dynamic simulations and for integrating Newton's equations [47] and [48]).

The idea of this method for the problem in Eq. (5.21) is very simple. Assuming that the function f is sufficiently regular, the Taylor formula, in a neighborhood of the point (x_0, t_0), gives

$$f(x,t) \approx f(x_0, t_0) + \Sigma_{i+j=1}^{n} a_{i,j}(x - x_0)^i (t - t_0)^j =: f_n(x,t) \tag{5.24}$$

Next, we consider the family of problems indexed by parameter λ

$$x'(t) = \lambda f_n(x,t), \quad x(t_0) = x_0, \tag{5.25}$$

where λ is some formal parameter and its value is set to 1 afterwards. It is natural to look for solution of Eq. (5.25) in the form

$$x_n(t) = x_0 + \Sigma_{k=1}^{n} \lambda^k \varphi_k(s), \tag{5.26}$$

where $s = t - t_0$ and $\varphi_k(s)$ are unknown functions with the property that $\varphi_k(0) = 0$. Putting Eq. (5.26) into Eq. (5.25) and next comparing the terms of the same order in λ in Eq. (5.25), we obtain the set of the differential equations for the unknown function φ_k of the form

$$\varphi_1'(s) = f(x_0, t_0) + \Sigma_{j=1}^{n} a_{0,j} s^j, \tag{5.27}$$

$$\varphi_2'(s) = (\Sigma_{j=1}^{n-1} a_{1,j} s^j) \varphi_1(s), \tag{5.28}$$

$$\varphi_3'(s) = (\Sigma_{j=1}^{n-2} a_{2,j} s^j) \varphi_2(s) + \Sigma_{j=1}^{n-1} a_{1,j} s^j \varphi_1^2(s), \tag{5.29}$$

and so on. The expressions for φ_k are obtained by integrating the above equations for φ_k' with respect to s, [47] and [48].

The labor involved in such a procedure can be efficiently handled by modern symbolic manipulation packages like *Mathematica* or *Mapple.* However, when instead of Eq. (5.25) we consider a system of differential equations then it is much more convenient to use a standarized programming language, *e.g.* *C++* together with the *GiNaC* library.

Note that the solutions of our system of equations with initial data $\varphi_k(0) = 0$ are polynomials, hence

$$x_n(t) = \Sigma_{k=1}^{n} \varphi_k(s) + x_0 \tag{5.30}$$

is a polynomial also. This polynomial is used in our algorithm; for a given accuracy $\varepsilon > 0$ and a natural number n we look for the interval $[0, h_1]$ such that

$$|x_n(t_0 + h) - x_{n-l}(t_0 + h)| < \varepsilon \quad \text{for} \quad h \le h_1 \quad \text{and} \quad l > 0. \tag{5.31}$$

Once ε has a given value, h_1 does not play the role of finite difference Δt which is approximating differential dt in finite-difference methods, but in many applications the resulting h_1 can take a large value. In practice, to get a satisfactory result it is enough to check the condition Eq. (5.31) for $n = 3$ and $l = 1, 2$. In this way we have, on the interval $[t_0, t_0 + h_1]$, an approximation of the solution $x^1(t)$ of the problem.

Next we continue our procedure starting at $t_0 + h_1$ and we obtain the function $x^2(t)$ defined on some interval $[t_0+h_1, t_0+h_1+h_2]$, which approximates on this interval the solution of the problem Eq. (5.21). Continuing this procedure we can find the solution for arbitrary interval $[0, T]$.

In the particular example of the Malthus-Verhust equation Eq. (5.4) the problem in Eq. (5.25) can be written as follows:

$$\frac{dN(t)}{dt} = \lambda(rN(t) - mN^2(t)), \tag{5.32}$$

where λ is some formal parameter. If $N(t_0) = N_0$ the expressions $\varphi_1(s)$, $\varphi_2(s)$, and $\varphi_3(s)$ in Eqs. (5.27) – (5.29) take the following form:

$$\varphi_1(s) = sN_0(r - mN_0), \tag{5.33}$$

$$\varphi_2(s) = -\frac{1}{2}s^2N_0(3mrN_0 - r^2 - 2m^2N_0^2), \tag{5.34}$$

$$\varphi_3(s) = -\frac{1}{6}s^3N_0(6m^3N_0^3 - r^3 - 12m^2rN_0^2 + 7mr^2N_0), \tag{5.35}$$

and the approximate solution which corresponds to polynomial of degree three in λ (we put $\lambda = 1$) is the following:

$$N_3(t) = N_0 + \varphi_1(s) + \varphi_2(s) + \varphi_3(s). \tag{5.36}$$

This polynomial together with a polynomial of a smaller degree, *e.g.*, $N_2(t) = N_0+\varphi_1(s)+\varphi_2(s)$ can be used as a accuracy criterion in Eq. (5.31). In Fig. 5.1, a few solutions obtained with the help of this method are presented for a given accuracy $\varepsilon = 10^{-30}$ but when the degree of the polynomial in λ is equal to 9. The results in practice are numerically the same as the exact solution Eq. (5.16). The logistic differential equation has very simple structure and in this particular case the degree of the polynomial in s is the same as the corresponding polynomial in λ. Generally it is not true.

A similar procedure can be used for the problem Eq. (5.22). The first step is to write $f(x(t), x(t-\tau))$ in the form

$$f(x,y) \approx f(x_0,x_0) + \Sigma_{i+j=1}^{N} a_i (x-x_0)^i (y-x_0)^j =: f_N(x,y), \qquad (5.37)$$

which approximates f in a neighborhood of the point (x_0, x_0). As in the previous case we consider the family of problems

$$x'(t) = \lambda f_N(x(t), x(t-\tau)), \quad x(t) = x_0(t), \text{ for } t \in [-\tau, 0]. \qquad (5.38)$$

For our purposes we assume that the initial function $x_0(t)$ is a polynomial. If not, we can approximate $x_0(t)$ by a polynomial with arbitrary accuracy. Using this assumption we can write, on the interval $[0, \tau]$, the problem Eq. (5.38) in the form Eq. (5.21), and to solve it we can use the procedure described above. In this way we get the solution on the interval $[0, \tau]$. Note that the approximate solution on the interval $[\tau, \tau + h]$ depends not only on $x_n(\tau)$ but also on the value of $x_n(t)$ on $[0, h]$. In our case on this interval $x_n(t)$ is a polynomial, hence we are able to repeat our procedure to define $x_n(t)$ on $[\tau, \tau + h]$, *i.e.* instead of $x(t-\tau)$ in Eq. (5.38) we put a formal expression in Eq. (5.26), where φ_k on $[0, h]$ are known from the first step in $[0, \tau]$ (Fig. 5.6). Continuing this procedure we can find the solution on arbitrary interval.

The following example shows what this memory dependence of the solutions of DDE looks like. To this aim, we consider the logistic equation Eq. (5.7) for $\Omega = r/m$ where the formal indexing parameter λ has been introduced:

$$\frac{dN(t)}{dt} = \lambda N(t)\left(r - mN(t-\tau)\right), \quad N(t) = N_0(t) \text{ for } t \in [-\tau, 0]. \qquad (5.39)$$

For simplicity, let $N_0(t) \equiv X_0$ in $[-\tau, 0]$ and let $N(0) = N_0 = X_0$ where X_0 is a constant. Then, in the interval $[0, \tau]$ the expressions $\varphi_1(s)$, $\varphi_2(s)$, and $\varphi_3(s)$ in Eqs. (5.27) - (5.29) after integrating them with respect to s take the following form:

$$\varphi_1(s) = N_0 s(r - mX_0), \qquad (5.40)$$

$$\varphi_2(s) = N_0 s^2 \left(\frac{1}{2} r^2 + X_0\left(\frac{1}{2} X_0 m^2 - mr\right)\right), \qquad (5.41)$$

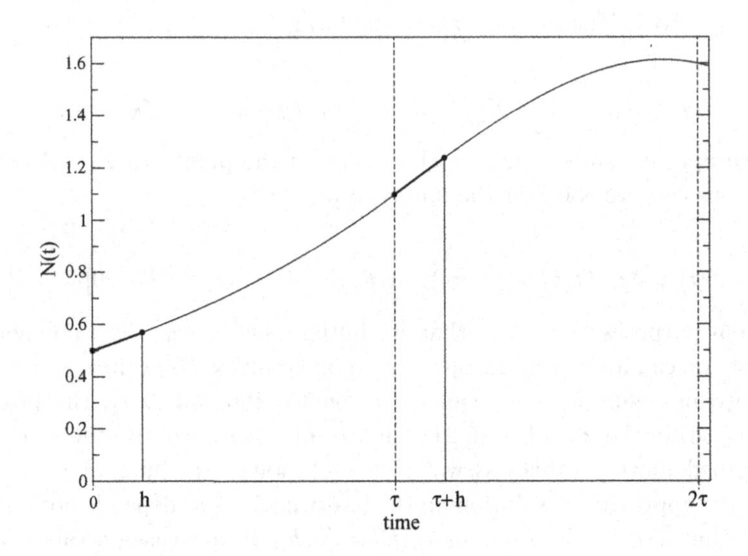

Fig. 5.6. The relationship between the solution of DDE in the successive time intervals $[0,\tau]$ and $[\tau, 2\tau]$. In order to calculate the solution on the interval $[\tau, \tau+h]$ it is necessary to know the solution on the interval $[0,h]$.

$$\varphi_3(s) = N_0 s^3 (\frac{1}{6}r^3 - \frac{1}{2}X_0(mr^2 + \frac{1}{3}X_0^2 m^3 - X_0 m^2 r)) \tag{5.42}$$

Notice that although N_0 is initially equal to X_0 it changes its value every time the accuracy condition in Eq. (5.31) fails whereas X_0 is a constant. It is the reason for keeping different notation for X_0 and N_0. In this case, the approximate solution which corresponds to polynomial of degree three in λ (we put $\lambda = 1$) is the following:

$$N_3(t) = N_0 + \varphi_1(s) + \varphi_2(s) + \varphi_3(s) = N_3(t; X_0, N_0). \tag{5.43}$$

However, in the interval $[\tau, 2\tau]$ we have a new initial condition $N_3(\tau) = N_1$ and therefore different expressions for $\varphi_1(s), \varphi_2(s)$, and $\varphi_3(s)$:

$$\varphi_1(s) = N_1 s(r - mN_0), \tag{5.44}$$

$$\varphi_2(s) = N_1 s^2 (\frac{1}{2}(N_0 m(-3r + m(N_0 + X_0)) + r^2)), \tag{5.45}$$

$$\varphi_3(s) = N_1 s^3 (\frac{1}{6}r^3 + \frac{5}{6}X_0 m^2 N_0 r - \frac{1}{2}X_0 m^3 N_0^2 - \frac{1}{6}X_0^2 m^3 N_0 + m^2 N_0^2 r - \frac{7}{6}m N_0 r^2 - \frac{1}{6}m^3 N_0^3). \quad (5.46)$$

Then, in $[\tau, 2\tau]$ the approximate solution which corresponds to polynomial of degree three in λ (we put $\lambda = 1$) is the following:

$$N_3(t) = N_1 + \varphi_1(s) + \varphi_2(s) + \varphi_3(s) = N_3(t; X_0, N_0, N_1). \quad (5.47)$$

The succeeding steps corresponding to time intervals $[2\tau, 3\tau]$, $[3\tau, 4\tau]$ etc. proceed in the same manner. With the help of the method using polynomials in formal parameter λ we have analyzed numerically all population differential equations, in particular Eqs. (5.12) for which the example solution is presented in Fig. (5.2). In this case the polynomials of degree 5 in λ were approximating $N_1(t)$ and $N_2(t)$.

The presented numerical method based on polynomials indexed with some formal parameter λ has been applied to DDE for the first time in this article. Furthermore, since the method is not a finite-difference method it does not introduce truncation errors connected with the Taylor series expansion in terms of the powers of the differences Δt which can introduce computer artifacts in non-linear problems. An example can be finite-difference representation of the logistic equation by the Runge-Kutta algorithms [49].

References

[1] V. Volterra, *Théorie mathématique de la lutte pour la vie.* (Gauthier-Villars, Paris, 1931).

[2] L. E. Reichl, *A Modern Course in Statistical Physics*, pp. 641-644. (Edward Arnold (Publishers) LTD, 1987).

[3] M. Abramowitz and I. A. Stegun, *Handbook of Mathematical Functions With Formulas, Graphs, and Mathematical Tables*, National Bureau of Standards, Applied Mathematics Series 55, p. 68, (1972).

[4] G. E. Hutchinson, Circular casual in ecology, *Ann. N. Y. Acad. Sci.* **50**, 221-246, (1948).

[5] L. Ramon, L. Lange and H. H. Hecht, The mechanism of Cheyne-Stokes respiration, *J. Clin. Invest.* **41**, 42-52, (1962). ISSN 0021-9738.

[6] W. W. Murdoch, C. J. Briggs, and R. M. Nisbet, *Consumer-Resource Dynamics.* (Princeton University Press, Princeton, NJ, USA, 2003).

[7] A. C. Guyton, J. W. Crowell, J. W. Moore, Basic Oscillating Mechanism of Cheyne-Stokes Breathing, *Am. J. Physiol.* **187**, 395-398, (1956). ISSN 0002-9513.

[8] Ch. T. H. Baker, G. A. Bocharov and F. A. Rihan, *A Report on the Use of Delay Differential Equations in Numerical Modelling in the Biosciences*, Manchester Centre for Computational Mathematics: Numerical Analysis Report No. 343, (1999). ISSN 1360-1725.

[9] M. Mackey and L. Glass, Oscillation and chaos in physiological control systems, *Science*, **197**(4300), 287-289, (1977). ISSN 0036-8075.

[10] M. Wazewska-Czyzewska, A. Lasota, Mathematical problems of the dynamics of a system of red blood cells, *Mat. Stos.* **6**, 23-40, (1976) (in Polish).

[11] Y. Zhao, M. Yamamoto, M. Munakata, M. Nakao, N. Katayama, Investigation of the time delay between variations in heart rate and blood pressure, *Med. Biol. Eng. Comput.* **37**, 344-347, (1999). ISSN 0140-0118.

[12] G. A. Bocharov, F. A. Rihan, Numerical modelling in bioscences using delay differential equations, *J. Comput. Appl. Math.* **125**, 183-199, (2000). ISSN 0377-0427.

[13] F. H. Wood and M. A. Foot, Graphical analysis of lag in population reaction to environmental change, *New Zeal. J. Ecol.* **4**, 45-51, (1981). ISSN 0110-6465.

[14] R. B. Banks, *Growth and Diffusion Phenomena.* Mathematical Frameworks and Applications in Texts in Applied Mathematics, vol.14, Springer, Berlin, (1994).

[15] I. Györi and F. Hartung, On the Exponential Stability of a State-Dependent Delay Equation, *Acta Sci. Math. (Szeged)* **66**, 87-100, (2000). ISSN 0001-6969.

[16] D. C. Gazis, E. W. Montroll, J. E. Ryniker, Age-specific, Deterministic Model of Predator-Prey Populations: Application to Isle Royale, *IBM J. Res. Develop.* **17**, 47-53, (1973). ISSN 0018-8646.

[17] K. E. F. Watt, A Mathematical Model for the Effect of Densities of Attacked and Attacking Species on the Number Attacked, *Canadian Entomology* **91**, 129, (1959).

[18] R. Gerami and M. R. Ejtehadi, A history-dependent stochastic predator-prey model: Chaos and its elimination, *Eur. Phys. J. B* **13**, 601-606, (2000). ISSN 1434-6028.

[19] V. Rai, W. M. Schaffer, Chaos in ecology, *Chaos Solit. Fract.* **12**, 197-203, (2001). ISSN 0960-0779.

[20] J. Vandermeer, L. Stone, B. Blasius, Categories of chaos and fractal basin boundaries in forced predator-prey models, *Chaos Solit. Fract.* **12**, 265-276, (2001). ISSN 0960-0779.

[21] R. M. May, Biological Populations with Nonoverlapping Generations: Stable Points, Stable Cycles, and Chaos. *Science* **186**, 645-647, (1974). ISSN 00036-8075.

[22] R.M. May, Simple mathematical models with very complicated dynamics, *Nature* **261**, 459-467, (1976). ISSN 0028-0836.

[23] D. Chowdhury and D. Stauffer, A food-web based unified model of "macro"- and "micro-" evolution, *Phys. Rev. E* **68**, 041901, (2003). ISSN 1539-3755.

[24] Ch. J. Harvey, S. P. Cox, T. E. Essington, S. Hansson, and J. F. Kitchell, An ecosystem model of food web and fisheries interactions in the Baltic Sea,

ICES J. Mar. Sci. **60**, 939-950, (2003). ISSN 1054-3139.
[25] Th. Hance, G. Van Impe, The influence of initial age structure on predator-prey interaction, *Ecol. Model.* **114**, 195-211, (1999). ISSN 0304-3800.
[26] W. Wang and L. Chen, A predator-prey system with stage-structure for predator, *Comput. Math. with Appl.* **33**, 83-91, (1997). ISSN 0898-1221.
[27] S. Liu, L. Chen, G. Luo, and Y. Jiang, Asymptotic behaviors of competitive Lotka-Volterra system with stage structure, *J. Math. Anal. Appl.* **271**, 124-138, (2002). ISSN 0022-247X.
[28] X. Zhang, L. Chen, and A. U. Neumann, The stage-structured predator-prey model and optimal harvesting policy, *Math. Biosci.* **168**, 201-210, (2000). ISSN 0025-5564.
[29] L. Cai, X. Li, X. Song, and J. You, Permanence and Stability of an Age-Structured Prey-Predator System with Delays. Discrete Dynamics in Nature and Society, **2007**, 1-15, (2007). ISSN 1026-0226.
[30] T. S. Ray, L. Moseley, N. Jan, A Predator-Prey Model with Genetics: Transition to a Self-Organized Critical State, *Int. J. Mod. Phys. C* **9**, 701, (1998). ISSN 0129-1831.
[31] T. J. P. Penna, A bit-string model for biological aging, *J. Stat. Phys.* **78**, 1629-1633, (1995). ISSN 0022-4715.
[32] S. Moss de Oliveira, P. M. C. de Oliveira and D. Stauffer, *Evolution, Money, War, and Computers*, (Teubner, Stuttgart-Leipzig, 1999).
[33] T. J. P. Penna, D. Stauffer, Bit-string ageing model and German population, *Z. Phys. B Con. Mat.* **101**, 469-470, 1996. ISSN 0722-3277.
[34] K. Bońkowska, S. Szymczak and S. Cebrat, Microscopic modeling the demographic changes, *Int. J. Mod. Phys. C* **17**, 1477-1487, (2006). ISSN 0129-1831.
[35] M. Kowalczuk, A. Łaszkiewicz, M. Dudkiewicz, P. Mackiewicz, D. Mackiewicz, N. Polak, K. Smolarczyk, J. Banaszak, M.R. Dudek and S. Cebrat, Using Monte-Carlo simulations for the gene and genome evolution, *Trends Stat. Phys.* **4**, 29-44, (2004). ISSN 0972-480X.
[36] F. Tria, M. Lä, L. Peliti and S. Franz, A minimal stochastic model for influenza evolution, *J. Stat. Mech.* P07008, (2005). ISSN 1742-5468.
[37] P. D. Turney, A Simple Model of Unbounded Evolutionary Versatility as a Largest-Scale Trend in Organismal Evolution, *Artif. Life* **6**, 109-128, (2000). ISSN 1064-5462.
[38] Y. Murase, T. Shimada, and N. Ito, Effects of stochastic population fluctuations in two models of biological macroevolution, arXiv:0803.137v1 [q-bio.PE], (2008).
[39] V. Sevim and P. A. Rikvold, Effects of correlated interactions in a biological coevolution model with individual-based dynamics. *J. Phys. A: Math. Gen.* **38**, 9475-9489, (2005). ISSN 0305-4470.
[40] P. Cortés, J. M. GarcÃŋía, J. Muñuzuri, L. Onieva, Viral systems: A new bio-inspired optimisation approach, *Comput. Oper. Res.* **35**, 2840-2860, (2008). ISSN 0305-0548.
[41] F. Bagnolia, C. Guardianic, A microscopic model of evolution of recombination, *Physica A* **347**, 489-533, (2005). ISSN 0378-4371.

[42] M. R. Dudek, Lotka-Volterra Population Model of Genetic Evolution, *Commun. Comput. Phys.* **2**, 1174-1183 (2007). ISSN 1815-2406.

[43] E. M. Jones, D. Paetkau, E. Geffen, C. Moritz, Genetic diversity and population structure of Tasmanian devils, the largest marsupial carnivore, *Mol. Ecol.* **13**(8), 2197-2209, (2004). ISSN 0962-1083.

[44] M. E. Jones, P. J. Jarman, C. M. Lees, H. Hesterman, R. K. Hamede, N. J. Mooney, D. Mann, Ch. E. Pukk, J. Bergfeld, and H. McCallum, Conservation Management of Tasmanian Devils in the Context of an Emerging, Extinction-threatening Disease: Devil Facial Tumor Disease, *EcoHealth* **4**, 326-337, (2007). ISSN 1612-9202.

[45] A. Duda, A. Nowicka, P. Dys, and M.R. Dudek, Effect of Inherited Genetic Information on Stochastic Predator-Prey Model, *Int. J. Mod. Phys. C* **11**, 1527-1538, (2000). ISSN 0129-1831.

[46] W. H. Press, S. A. Teukolsky, W. T. Vetterling, B. P. Flannery, *Numerical Recipes in C: the art of scientific computing* - 2nd ed., (Cambridge University Press, 1992).

[47] M. R. Dudek and T. Nadzieja, *Molecular dynamics simulations through symbolic programming*, *Int. J. Mod. Phys. C.* **16**, 413-425, (2005). ISSN 0129-1831.

[48] B. Brzostowski, M. R. Dudek, B. Grabiec, and T. Nadzieja, *Non-finite-difference algorithm for integrating Newton's motion equations*, *Phys. Stat. Sol. (b)* **244**, 851-858, (2007). ISSN 0370-1972.

[49] J. H. E. Cartwright and O. Piro, *The dynamics of Rune-Kutta methods*, *Int. J. Bifurc. Chaos* **2**, 427-449, (1992). ISSN 0218-1274.

Chapter 6

Computational modeling of evolution: ecosystems and language

A. Lipowski[1] and D. Lipowska[2]

[1]*Faculty of Physics, Adam Mickiewicz University, 61-614 Poznań, Poland, lipowski@amu.edu.pl*

[2] *Institute of Linguistics, Adam Mickiewicz University, Poznań, Poland, lipowska@amu.edu.pl*

Recently, computational modeling became a very important research tool that enables us to study problems that for decades evaded scientific analysis. Evolutionary systems are certainly examples of such problems: they are composed of many units that might reproduce, diffuse, mutate, die, or in some cases for example communicate. These processes might be of some adaptive value, they influence each other and occur on various time scales. That is why such systems are so difficult to study. In this paper we briefly review some computational approaches, as well as our contributions, to the evolution of ecosystems and language. We start from Lotka-Volterra equations and the modeling of simple two-species prey-predator systems. Such systems are canonical example for studying oscillatory behavior in competitive populations. Then we describe various approaches to study long-term evolution of multi-species ecosystems. We emphasize the need to use models that take into account both ecological and evolutionary processes. Recently we introduced a simple model of this kind, and its behavior is briefly summarized. In this multi-species prey-predator system, competition of predators for preys and space results in evolutionary cycling. We suggest that such a behavior of the model might correspond to long-term periodic changes of the biodiversity of the Earth ecosystem as predicted by Raup and Sepkoski. Finally, we address the problem of the emergence and development of language. It is becoming more and more evident that any theory of language origin and development must be consistent with darwinian principles of evolution. Consequently, a number of techniques developed for modeling evolution of complex ecosystems are being applied to the problem of language. We briefly review some of these approaches. We also discuss the behavior of a recently introduced evolutionary version of the naming-game model. In this model communicating agents reach linguistic coherence via a bio-linguistic transition which is due to the coupling of evolutionary and linguistic abilities of agents.

Contents

6.1. Introduction

Evolution is a fundamental property of life. It consists of two basic and in a sense opposing ingredients. Mutations and crossing-over are forces that increase variability of organisms and eventually lead to the formation of new brands and species. Against these processes acts selection and due to limited resources only best adapted organisms survive and pass their genetic material to the next generation.

Evolutionary forces are operating since the very early emergence of life. To large extent they shaped the complicated pattern of past and present speciation and extinction processes. Even qualitative understanding of the dynamics that governs these processes is a very challenging problem. Eldredge and Gould [1, 2] noted that palaeontological data show that intensity of speciation and extinction processes varied throughout the life history and periods of evolutionary stagnation were interrupted with bursts of activity (punctuated equilibrium). Such a pattern, at least for a physicist, resembles the behavior of a system at a critical point. Indeed, similarly to critical systems, some palaeontological data can be also described with power laws [3]. Following the work of Bak and Sneppen [4], a lot of models that try to explain why an Earth ecosystem might be considered as a critical system, were examined [3]. But the quality of palaeontological data does not allow for definitive statements and some alternative interpretations were also proposed. In particular, an interesting conjecture was made by Raup and Sepkoski who suggested that the pattern of bursts of extinctions is actually periodic in time with a period of approximately 26 mln. years [5]. Despite an intensive research it is not clear what a factor could induce such a periodicity, and certainly, it would be desirable to understand the main macroevolutionary characteristics of Earth's ecosystem.

But it is not only intensity of extinction and speciation processes that is of some interest. In the evolution of life one can distinguish several radical changes that had a dramatic consequences and lead to the emergence of new levels of complexity [6]. As examples of such changes Maynard-Smith and Szathmáry list invention of genetic code, transition from Prokaryotes to Eukaryotes, or appearance of colonies. Certainly, such changes had a tremendous impact on the evolution and the current state of the Earth's ecosystem. As a last major transition of this kind, Maynard-Smith and Szathmáry mention the emergence of human societies and language. To large extent this transition was induced by cultural interactions. Such interactions when coupled to evolutionary processes lead to the immensely complex system - human society. According to Dawkins [7], cultural interactions open a new route to the evolution that is no longer restricted to living forms. Examples of such forms, called memes, include songs (not only human), well known sentences and expressions, fashion or architecture styles. As a matter of fact, events listed as major transitions have an interesting feature in common - they created new mechanisms of information transmission. Language certainly enables transmission and storage of very complex cultural information. Its emergence enormously speeded up the information transfer between generations. Before that, for nearly four billion years of life on Earth, the only information that could be used for evolutionary purposes was encoded in the genome. With language vast amount of information can be exchanged between humans and passed on to subsequent generations. Most likely it was the invention of language that enormously speeded up the the evolution of our civilization and made humans in a sense a unique species. Although the emergence and subsequent evolution of language had tremendous influence, this process is still to large extent mysterious and is considered as one of the most difficult problems in science [8].

Computer modeling seems to be a very promising technique to study complex systems like ecosystems or language. In the present chapter we briefly review such an approach and present our results in this field. In Sec. 6.2 we briefly discuss population dynamics of simple two-species prey-predator systems and classical approaches in this field based on Lotka-Volterra equations. We also argue that it is desirable to use an alternative approach, the so-called individual based modeling. An example of such a model is described in Sec. 6.3. In this section we discuss results of numerical simulations of the model concerning especially the oscillatory behavior.

Processes in simple ecosystems with constant number of non-evolvable species (as described in Sec. 6.2) take place on ecological time scale. To describe real, *i.e.*, complex, ecosystems we have to take into account also evolutionary processes, such as speciations or extinctions. Such processes operate on the so-called evolutionary time scale. Such a time scale was usually regarded as much longer than ecological time scale, however, there is a number of examples that show that they are comparable [9, 10]. In Sec. 6.4 we briefly review models used to study complex ecosystems. In particular, we emphasize the need to construct models that would take into account both ecological and evolutionary processes. In Sec. 6.5 we examine one of such models which is a multi-species generalization of prey-predator model studied in Sec. 6.3. Investigating extinction of species we show that their intensity changes periodically in time. The period of such oscillations is set by the mutation rate of the model. Since evolutionary changes are rather slow, we expect that such oscillations in the real ecosystem would have very long periodicity. We suggest that such a behavior agrees with the conjecture of Raup and Sepkoski, but more detailed analysis of the predictions of our model would be desirable. In the final part of this section we suggest that our model might provide an insight into a much different problem. Namely, we attempt to explain the uniqueness of the coding mechanism of living cells as contrasted with the multispecies structure of present-day ecosystems. Apparently, at the early stage of life a primitive replicator happened to invent the universal code that was so effective that it spread over the entire ecosystem. However, at a certain point such a single-species ecosystem become unstable and was replaced by a multi-species ecosystem. In our model, upon changing a control parameter, a similar transition (between single- and multi-species ecosystems) takes place and we argue that it might be analogous to the early-life transition.

In Sec. 6.6 we review computational studies on language evolution. An important class of models is based on the so-called naming game introduced by Steels [11]. Recently, we examined an evolutionary version of this model and showed that coupling of evolutionary and linguistic interactions leads to some interesting effects [12]. Namely, for sufficiently large intensity of linguistic interactions, there appears an evolutionary pressure that rapidly increases linguistic abilities and the model undergoes an abrupt biolinguistic transition. In such a way communicating agents establish a common vocabulary and the model reaches the so-called linguistic coherence. Our model incorporates both learning and evolution. Interaction of these two factors, known as a Baldwin effect [13], is recently intensively studied also in the context of language evolution [14]. Discussion of the Baldwin effect and related properties in our model is also presented in Sec. 6.6. We conclude in Sec. 6.7.

6.2. Coarse-grained versus individual-based modeling of an ecosystem

Population dynamics provides the basis of the modeling of the ecosystem. Pierre Verhulst, regarded as its founding father, noticed that due to the finite environmental capacity the unlimited growth of the population predicted by the linear growth equation is unrealistic. Consequently, Verhulst proposed that the time evolution of the population should be described by the following equation [15]

$$\frac{dx}{dt} = kx(1 - x/x_M) \tag{6.1}$$

where k is the growth rate, x is the size of the population, and x_M is the environmental capacity. Equation (6.1), that is called logistic equation, found numerous applications in demographic studies. However, to describe any realistic ecosystem one should consider more, possibly interacting, populations. A step in this direction was made by Lotka who examined a simple autocatalytic reaction model [16]. His work was followed by Volterra who wrote down essentially the same set of ordinary differential equations (ODE) studying the statistics of fish catches [17]. Lotka-Volterra equations for two interacting populations of preys (x) and predators (y) can be written as

$$\begin{aligned}\frac{dx}{dt} &= x(a_1 - a_2 y)\\ \frac{dy}{dt} &= y(-a_3 + a_4 x)\end{aligned} \tag{6.2}$$

where a_1, a_2, a_3, and a_4 are some positive constants. Although Eqs. (6.2) constitute the canonical model to study periodic oscillations in competitive systems [18], they were also criticized on various grounds. For example their solution depends on the initial condition, and the very form of Eqs. (6.2) is structurally unstable. It means that their small modification (with *e.g.*, higher order terms like x^2y) will typically destroy oscillatory behavior. Although there are some ODE models where such a limit cycle behavior is more stable [19], an important feature of any realistic system is missing in Eqs. (6.1) or (6.2). Namely, they neglect spatial heterogeneities. The simplest way to take them into account would be to consider x and y as spatially dependent quantities and then to replace Eqs. (6.1) or (6.2) with their partial differential analogs. After such a modification Eq. (6.1) becomes the famous Fisher equation, that in the one-dimensional case has the form

$$\frac{dx}{dt} = kx(1 - x/x_M) + D\frac{\partial^2 x}{\partial l^2} \tag{6.3}$$

where l is the spatial coordinate, D is the diffusion constant and $x = x(t,l)$ depends on t and l. Various extensions of (6.3), that are called sometime reaction-diffusion models, were also intensively studied in ecological contexts [18, 20].

Although description in terms of partial differential equations takes into account some of spatial heterogeneities, it is still based on the coarse-grained quantities like $x(t,l)$ and that means that it is essentially of the mean-field nature. Moreover, kinetic coefficients (k, D, x_M, a_1, a_2, a_3, $a_4, \ldots$) that enter such equations are usually difficult to determine from ecological data. Similar problems appear in alternative approaches to spatially extended ecological models based on coupled-map lattices [21] or integrodifference equations [22]. It is thus worth to pursue an alternative approach, the so-called individual based modeling, where to some extent stochastic rules, mimicking realistic processes like death, breeding or movement, are formulated at the level of individual organisms. Models of ecosystems formulated within such an approach are particularly suited for numerical computations and resemble some nonequilibrium statistical mechanics models. Such a similarity is very valuable since the behavior of ecological systems can be put in a wider perspective.

6.3. Lattice prey-predator models

To simplify calculations individual-based models of prey-predator systems are usually formulated on a cartesian d-dimensional lattice of the linear size N. One can define dynamics of such models in various ways, but to provide a detailed example we present rules used in some of our previous works [23, 24]. In our model on each site i of a lattice there is a four-state variable $\epsilon_i = 0, 1, 2, 3$ which corresponds to the site being empty ($\epsilon_i = 0$), occupied by a prey ($\epsilon_i = 1$), occupied by a predator ($\epsilon_i = 2$) or occupied by a prey and a predator ($\epsilon_i = 3$). Its dynamics has one control parameter r ($0 \leq r \leq 1$) and is specified as follows:

- Choose a site at random.
- With the probability r update a prey at the chosen site, provided that there is one (*i.e.*, $\epsilon = 1$ or 3); otherwise do nothing. Provided that at least one neighbor of the chosen site is not occupied by a prey (*i.e.*, $\epsilon = 0$ or 2), the prey (which is to be updated) produces one offspring and places it on the empty neighboring site (if there are more empty sites, one of them is chosen randomly). Otherwise (*i.e.*, when there is a prey on each neighboring site) the prey does not breed (due to overcrowding).
- With the probability $1 - r$ update a predator at the chosen site, provided that there is one (*i.e.*, $\epsilon = 2$ or 3). Provided that the chosen site is occupied by a predator but is not occupied by a prey ($\epsilon = 2$), the predator dies (of hunger). If there is a prey on that site (*i.e.*, $\epsilon = 3$), the predator survives and consumes the prey from the site it occupies. If there is at least one neighboring site which is not occupied by a predator, the predator produces one offspring and places it on the empty site (chosen randomly when there are more such sites).

As neighboring sites, *i.e.*, sites where offspring can be placed, we usually consider the nearest neighbors, but taking into account further neighbors does not change the results qualitatively [25]. To characterize the behavior of the model let us introduce the densities of preys (x) and predators (y) defined as

$$x = \frac{1}{N^d} \sum_i (\delta_{\epsilon_i,1} + \delta_{\epsilon_i,3}), \quad y = \frac{1}{N^d} \sum_i (\delta_{\epsilon_i,2} + \delta_{\epsilon_i,3}), \tag{6.4}$$

where summation is over all N^d sites i and δ is KroneckerŠs δ-function. Of the main interest are actually averages $\langle x \rangle$ and $\langle y \rangle$, where averaging is over simulation time.

From the above rules it follows that the model has two absorbing states (*i.e.*, once the model enter such a state it remains trapped there for ever). The first one is filled with preys only ($x = 1$, $y = 0$) and the second one is empty ($x = 0$, $y = 0$). Simulations [23–25] show that for large enough r, both populations coexist and the model is in the active phase ($x > 0,\ y > 0$). When the update rate of preys r decreases, their number becomes to small to support predators. For sufficiently small r predators die out and the model quickly reaches the absorbing state where it is filled with preys (the empty absorbing state has a negligible probability of being reached by the model dynamics). The phase transition between active and absorbing phase was observed at positive r for $d = 1, 2$ and 3. At least for $d = 1$ (linear chain) Monte-Carlo simulations clearly show that the phase transition belongs to the directed percolation (DP) universality class [24] (see Fig. 6.1).

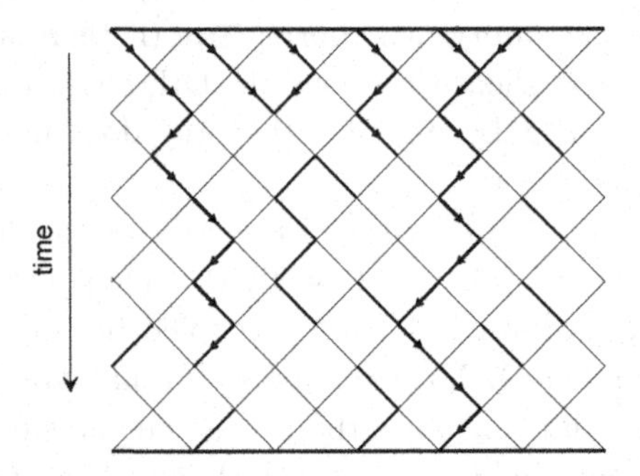

Fig. 6.1. In the directed percolation problem a fraction of bonds on a lattice is permeable (thick lines). From the top horizontal line water starts to flow downward through permeable bonds. Bonds that are reachable by water are shown with arrows. If at a certain (horizontal) level there would be no water, there would be no water below that level (absorbing state). When concentration of permeable bonds exceeds a certain threshold value a cluster of permeable bonds that spans from top to bottom is formed. At the threshold value the model is critical. A lot of dynamical models with a single absorbing state belong to the directed percolation universality class

Such a behavior is not surprising. There are by now convincing numerical and analytical arguments that various models possessing a single absorbing state generically belong to the DP universality class [26]. Moreover, models with multiple, but asymmetric absorbing states (such as *e.g.*, the model analyzed in this section) also belong to this universality class (models with multiple but symmetric absorbing states typically belong to some other universality classes, or undergo discontinuous phase transitions [27]). However, studying the critical behavior of models with absorbing-state phase transitions is not entirely straightforward. For finite systems (which is obviously the case in various simulational techniques) and close to the critical point, the model has a non-negligible probability of entering an absorbing state even when control parameters are such that the infinite system would remain in the active phase. Such a behavior sets a size-dependent timescale (*i.e.*, the lifetime of the active state) that severely affects simulations. A special technique, the so-called dynamical Monte-Carlo, is needed to obtain precise estimation of critical exponents for models of this kind [26, 28].

Of our main interest, however, is the oscillatory behavior of the model. To examine it, we measured the variances of the densities x and y as well as their Fourier transforms. Simulations show that for $d = 1$ and $d = 2$ in the limit $N \to \infty$ stochastic fluctuations wash out the oscillatory behavior and the variances of densities vanish. However, for $d = 3$ in the active phase and close to the absorbing transition, there is a range of r where oscillatory behavior survives in the limit $N \to \infty$.[a] Oscillations occur essentially for any initial conditions and their period only weakly depends on the parameter r.

It is the dimension of the lattice d that most likely plays an important role. Indeed, simulations show that for $d = 2$ models but with larger number of neighboring sites oscillations are again washed out in the limit $N \to \infty$ [25]. Such a result is in agreement with some arguments of Grinstein *et al.* [29] who related temporal periodic phases of noisy extended systems and smooth interfaces in growth models and concluded that oscillations might exist but only for $d > d_c = 2$. The (r, d) phase diagram of our model is sketched in Fig. 6.2.

To get an additional insight into the behavior of the model we can write mean-field equations that describe the time evolution of the densities x and y. Simple arguments [23] lead to the following set of equations

$$\begin{aligned} \frac{dx}{dt} &= rx(1 - x^w) - (1 - r)xy \\ \frac{dy}{dt} &= (1 - r)xy(1 - y^w) - (1 - r)y(1 - x) \end{aligned} \tag{6.5}$$

where w is the number of neighboring sites, as defined in the dynamical rules of the model (in most of our simulations neighboring sites were nearest neighbors and in such a case $w = 2d$). For example in the first equation of (6.5), the first term $(rx(1 - x^w))$ describes the growth rate of preys due

[a] Such a conclusion is based on the non-vanishing of variances of densities in this limit. Strictly speaking, based solely on such a behavior, one cannot exclude that this is *e.g.*, chaotic behavior that sets in. However, a pronounced peak in the Fourier transform of the time-dependent densities strongly supports the oscillatory interpretation of the numerical data.

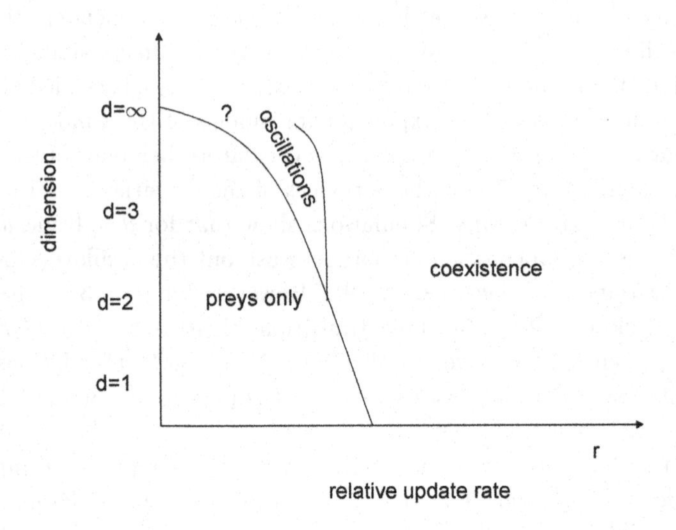

Fig. 6.2. The schematic phase diagram of the lattice prey-predator model. The phase transition separating preys-only and coexistence phases most likely belongs to the directed percolation universality class [24], but in more general models other transitions are possible [30]. It is not clear what is the fate of the oscillatory phase in $d > 3$ case, since the mean-field Eqs. (6.5) (that most likely correctly describe the model in large dimension) do not predict the oscillatory regime.

to updating a site with prey (rx) that happen to have at least one empty neighboring site ($(1 - x^w)$). The second term ($(1 - r)xy$) describes the decrease rate of preys due to an update of a site that happened to be a predator and that is also occupied by a prey. However, predictions of approximation (6.5) even qualitatively disagree with numerical simulations. In particular, in any dimension d the approximation (6.5) predicts that for any positive r there is no phase transition between active and absorbing phases. Moreover, within this approach there is no indication of the oscillatory phase, as observed in Monte-Carlo simulations for $d = 3$.

In the approximation (6.5) the probability that a site is occupied by a prey and predator is given as xy. This is of course only an approximation, and a much better scheme is obtained where this probability is considered as yet another variable (z), whose evolution follows from the dynamical rules of the model. In such a way we arrive at the following set of equations [25]

$$\begin{aligned}
\frac{dx}{dt} &= rx(1-x^w) - (1-r)z \\
\frac{dy}{dt} &= (1-r)z(1-y^w) - (1-r)(y-z) \\
\frac{dz}{dt} &= \frac{rx(1-x^w)(y-z)}{1-x} - \frac{(1-r)z(1+z-x-y)(1-y^w)}{1-y} - (1-r)zy^w
\end{aligned} \tag{6.6}$$

In the first term of the third equation of Eqs. (6.6) $x(1-x^w)$ gives the probability that the chosen site contains the prey and at least one of its neighbors does not. The factor $\frac{y-z}{1-x}$ gives the probability that the site chosen for reproduction of the prey is occupied by the predator only. The set of Eqs. (6.6) remains in a much better (than (6.5)) agreement with Monte-Carlo simulations [31]. In particular it predicts, oscillatory regime for $w \geq 4$. In Monte-Carlo simulations of models on cartesian lattices (we made simulations only for $d = 1, 2, 3$) oscillations appear only in the $d = 3$ case (*i.e.*, $w = 6$), and for $d = 2$ (*i.e.*, $w = 4$) these were only quasi-oscillations with the amplitude that vanishes in the limit $N \to \infty$.

It is interesting to ask what is the mechanism that triggers the emergence of finite-amplitude oscillations. Rosenfeld *et al.* suggested [32, 33] that oscillatory behavior in another lattice prey-predator model is induced by some kind of percolation transition (see Fig. 6.3). However, precise measurements of cluster properties in our model has shown that although some percolation transitions are indeed close to the onset of oscillatory regime, they clearly do not overlap with this onset [31]. Another proposal relates oscillations with some kind of stochastic resonance [23, 34]. Such a relation might be suggested by the mean-field approximation (6.5) that in fact describes a quasi-oscillatory dynamical system. Stochastic fluctuations, that are present in the lattice model but are neglected in the mean-field description, might be considered as a noise perturbing such a dynamical system. As shown by Gang *et al.* [35] due to stochastic resonance, in some low-dimensional autonomous dynamical systems noise might induce oscillatory behavior and one can expect that a similar scenario operates in lattice prey-predator systems.

There are also other lattice prey-predator models were similar oscillations were observed [38–43]. In a more general model, were predation and reproduction time scales are independent, a first-order phase transition might appear [30]. One can also mention that there are some important ecological problem that so far were not examined with individual based

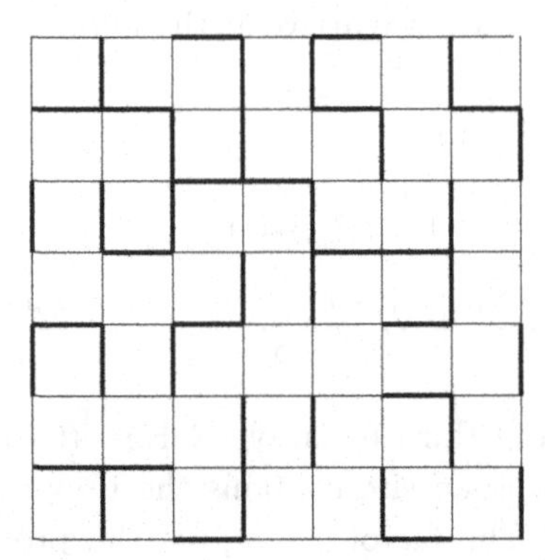

Fig. 6.3. In the bond percolation problem a fraction c of bonds is occupied (thick lines) [36, 37]. Neighboring occupied bonds form clusters. When c exceeds a certain threshold value an infinite (*i.e.*, spanning the entire lattice) cluster is formed. In a related problem of site percolation a fraction of sites is occupied and neighboring occupied sites form clusters. As suggested by Rosenfeld *et al.* [32, 33] oscillatory behavior might be induced by some kind of percolation transition *i.e.*, formation of an infinite cluster of preys or predators. Although the idea is appealing some calculations do not support it [31].

modelling but where such an approach might prove to be valuable. In this context one can mention various synchronization problems in spatially extended ecological systems [44] and in particular the Moran effect describing synchronization of populations exposed to common noise [45].

6.4. Modeling of complex ecosystems

Models that we discussed in the previous section describe rather simple ecosystems composed of few (two, three,...) species. Dynamics of such models implements basic ecological processes: reproduction, death, and in some cases also migration or aging. In such models changes of the populations takes place on a characteristic ecological time scale that is set by the dynamics of the models Typically, in real ecosystems this scale is of the order of years (for example, in a hare-lynx system oscillations with the period of approximately 10 years were identified [46]). But there are also some other than ecological processes. On the so-called evolutionary (or geological) time scale the entire species might die, change, or give rise to a new species. The evolutionary time scale is usually considered as much longer than ecological one [47]. As a result very often researchers constructed specific models directed toward either ecological or evolutionary processes.

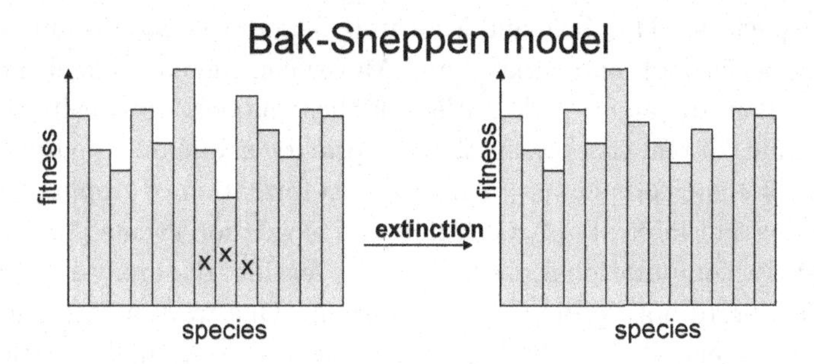

Fig. 6.4. The one-dimensional version of the Bak-Sneppen model. First, the species of lowest fitness is selected. Then this species and two of its neighbors (denoted by crosses) go extinct and are replaced by three new species with randomly selected fitness. Such a simple dynamics (no control parameters) drives the model into the critical state. The extremal dynamics of the Bak-Sneppen model was criticized on biological grounds, nevertheless this model drew considerable attention and is one of the main models of the self-organized criticality.

However, there are numerous examples showing that these time scales are not that much different and in some cases they are even comparable [9, 10]. Thus, such a separation of time scales is to some extent artificial and was used mainly for the ease of modeling (for a theoretical discussion of some related issues see *e.g.*, the paper by Khibnik and Kondrashov [48]). Actually, the complexity of real multi-species ecosystems and the difficulty to model them to some extent follow from the fact that these scales are not completely separated and ecological and evolutionary processes affect each other.

A model of multi-species ecosystem that tries to describe evolutionary processes and drew considerable attention especially in physicists community was introduced by Bak and Sneppen [4] (see Fig. 6.4). An interesting property of this model is its self-organized criticality. Namely, dynamics drives the model into the state where extinctions are strongly correlated (like in critical systems) and such a behavior resembles the punctuated equilibrium hypothesis of Eldredge and Gould [1, 2]. However, the Bak-Sneppen model has the dynamics that is operating at the level of species and refers to the (still controversial) notion of fitness. Thus, the model neglects ecological effects and despite rich and intriguing dynamics can be considered only as a toy model of an ecosystem. Nevertheless, the work of

Bak and Sneppen inspired other researchers to examine a number of models with species-level dynamics. For example, Vandevalle and Ausloos incorporated speciation [49], Solé and Manrubia introduced various interactions between species [50], and Amaral and Meyer considered some elements of the food-chain dynamics [51]. Although these models drastically simplify the dynamics of real ecosystems they do provide a valuable qualitative description of some complex problems such as formation of trophic levels or correlations and intensity of speciation and extinction events [3].

Recently, computational methods made feasible the analysis of models that incorporate both ecology and evolution. One way to construct such models is to generalize Lotka-Volterra equations to the multi-species case and to implement some speciation and extinction mechanism. Such an approach has already been developed [52–54], but it has similar drawbacks as original Lotka-Volterra model, namely it neglects spatial heterogeneities. In an alternative approach one uses individual-based dynamics and some models of multi-species ecosystems equipped with such a dynamics were examined [55]. A diagram that illustrates some types of models and their range of applicability is shown in Fig. 6.5.

6.5. Multispecies prey-predator model and periodicity of extinctions

In this section we describe the multi-species version of a lattice prey-predator model [56, 57]. Numerical simulations of the model show that the periodicity of mass extinctions, that was suggested by Raup and Sepkoski [5], might be a natural feature of the ecosystem's dynamics and not the result of a periodic external perturbation.

6.5.1. *Model*

Our model might be considered as a generalization of the two-species model described in Sec. 6.3. The model is defined on a d-dimensional cartesian lattice of the linear size N. Similarly to the two-species model one uses the four-state variables $\epsilon_i = 0, 1, 2$ or 3. In addition, each predator is characterized by its size m $(0 < m < 1)$ that determines its consumption rate and at the same time its strength when it competes with other predators. Only approximately the size m can be considered as related with physical size. Predators and preys evolve according to rules typical to such systems (*e.g.*, predators must eat preys to survive, preys and predators can breed provided

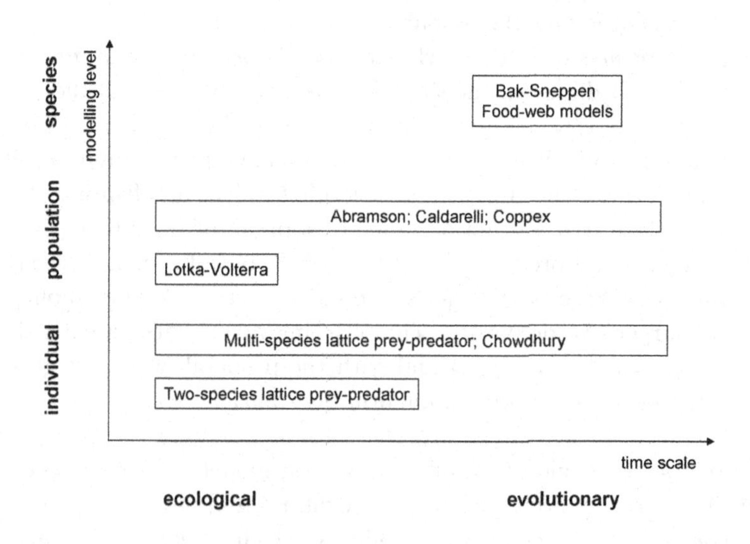

Fig. 6.5. Ecological and evolutionary aspects of the modeling of an ecosystem. The most coarse-grained models of ecosystem have dynamics operating at the level of species. Such models (*e.g.*, Bak-Sneppen [4], Food-web models [51]) neglect population level processes and describe ecosystem at the evolutionary time scale. Models originating from the Lotka-Volterra model use dynamics defined at the population level. Such models describe ecological processes in few-species ecosystems, but multi-species versions of Abramson [52], Caldarelli *et al.* [53] or Coppex *et al.* [54], that encompass also the evolutionary process were examined as well. Similar range of applicability have models with individual level dynamics (the model of Chowdhury *et al.* [55] neglects, however, the heterogeneities in spatial distribution of organisms).

that there is an empty site nearby, *etc.*). In addition, the relative update rate for preys and predators is specified by the parameter r $(0 < r < 1)$ and during breeding mutations are taking place with the probability p_{mut}. More detailed definition of the model dynamics is given below:

- Choose a site at random (the chosen site is denoted by i).
- Provided that i is occupied by a prey (*i.e.*, if $\epsilon_i = 1$ or $\epsilon_i = 3$) update the prey with the probability r. If at least one neighbor (say j) of the chosen site is not occupied by a prey (*i.e.*, $\epsilon_j = 0$ or $\epsilon_j = 2$), the prey at the site i produces an offspring and places it on an empty neighboring

site (if there are more empty sites, one of them is chosen randomly). Otherwise (*i.e.*, if there are no empty sites) the prey does not breed.

- Provided that i is occupied by a predator (*i.e.*, $\epsilon_i = 2$ or $\epsilon_i = 3$) update the predator with the probability $(1-r)m_i$, where m_i is the size of the predator at site i. If the chosen site i is occupied by a predator only ($\epsilon_i = 2$), it dies, *i.e.*, the site becomes empty ($\epsilon_i = 0$). If there is also a prey there ($\epsilon_i = 3$), the predator consumes the prey (*i.e.*, ϵ_i is set to 2) and if possible, it places an offspring at an empty neighboring site. For a predator of the size m_i it is possible to place an offspring at the site j provided that j is not occupied by a predator ($\epsilon_j = 0$ or $\epsilon_j = 1$) or is occupied by a predator ($\epsilon_j = 2$ or $\epsilon_j = 3$) but of a smaller size than m_i (in such a case the smaller-size predator is replaced by an offspring of the larger-size predator). The offspring inherits its parent's size with the probability $1 - p_{mut}$ and with the probability p_{mut} it gets a new size that is drawn from a uniform distribution.

At first sight one can think that such a model describes an ecosystem with two trophic levels (preys and predators) and only with predators being equipped with evolutionary abilities, which would be of course highly unrealistic. Let us notice, however, that expansion of predators sometimes proceeds at the expense of smaller-size predators. Thus, predators themselves are involved in prey-predator-like interactions. Perhaps it would be more appropriate to consider unmutable preys as a renewable (at a finite rate) source of, *e.g.*, energy, and predators as actual species involved in various prey-predator interactions and equipped with evolutionary abilities.

In the remaining part of this section we will describe a possible application of our model to the problem of mass extinctions and to the problem of multiplicity of species in the Earth ecosystem as contrasted with the uniqueness of the genetic code.

6.5.2. *Extinctions*

The suggestion that mass extinctions might be periodic in time was made by Raup and Sepkoski [5]. While analyzing fossil data, they noticed that during the last 250 My (million years) mass extinctions on Earth appeared more or less cyclically with a period of approximately 26My. Although their analysis was initially questioned [58], some other works confirmed Raup and Sepkoski's hypothesis [59–61]. The suggested large periodicity of mass extinctions turned out to be very difficult to explain. Indeed, 26My does not seem to match any of known Earth cycles and some researchers have been looking for more exotic explanations involving astronomical effects [62, 63], increased volcanic activity [64], or the Earth's magnetic field reversal [65]. So far, however, none of these proposals has been confirmed. One should also note that the most recent analysis of palaeontological data that span last 542My strongly supports the periodicity of mass extinctions albeit with a larger cycle of about 62My [66].

Lacking a firm evidence of an exogenous cause, one can ask whether the periodicity of extinctions might be explained without referring to such a factor. In Secs. 6.2 and 6.3 we already mentioned that periodic behavior of some prey-predator systems is not the result of periodic driving but rather a natural feature of their dynamics. However, the period of oscillations in such systems is determined by the growth and death rate coefficients of interacting species and is of the order of a few years rather than tens of millions. Consequently, if the periodicity of mass extinctions is to be explained within a model of interacting species, a different mechanism that generates long-period oscillations must be at work.

Such a mechanism might be at work in the multi-species model described in Sec. 6.5.1. Numerical simulations [56, 57] show that the model generates long-period evolutionary oscillations The period of these oscillations is determined by the inverse of the mutation rate and we argued that it should be several orders of magnitude longer than in the Lotka-Volterra oscillations. The mechanism that generates oscillations in our model can be briefly described as follows: A coevolution of predator species induced by the competition for food and space causes a gradual increase of their size. However, such an increase leads to the overpopulation of large predators and a shortage of preys. It is then followed by a depletion of large species and a subsequent return to the multi- species stage with mainly small species that again gradually increase their size and the cycle repeats. Numerical calculations for our model show that the longevity of a species depends on the evolutionary stage at which the species is created. A similar pattern has been observed in some palaeontological data [67] and, to our knowledge, the presented model is the first one that reproduces such a dependence. Let us notice that the oscillatory behavior in a prey-predator system that was also attributed to the coevolution has been already examined by Dieckmann *et al.* [68]. In their model, however, the number of species is kept constant and it cannot be applied to study extinctions. Moreover, the idea that an internal ecosystem dynamics might be partially responsible for the long-term periodicity in the fossil records was suggested by Stanley [69] and later examined by Plotnick and McKinney [70]. However, according to Stanley mass extinctions are triggered by external impacts. Their approximately equidistant separation is the result of a delayed recovery of the ecosystem. In our approach no external factor is needed to trigger such extinctions and sustain their approximate periodicity.

A gradual increase of size of species in our model recalls the Cope's rule that states that species tend to increase body size over geological time. This rule is not commonly accepted among paleontologists and evolutionists and was questioned on various grounds [71]. However, recent studies of fossil records of mammal species are consistent with this rule [72, 73]. Perhaps our model could suggests a way to obtain a theoretical justification of this rule.

Although very complicated, in principle, it should be possible to estimate the value of the mutation probability p_{mut} from the mutational properties of living species. Let us notice that in our model mutations produce an individual that might be substantially different from its parent. In Nature, this is typically the result of many cumulative mutations and thus we expect that p_{mut} is indeed a very small quantity. Actually, p_{mut} should be considered rather as a parameter related with the speed of morphological and speciation processes that are known to be typically very slow [47]. Perhaps a different version of the mutation mechanism where a new species would be only a small modification of its parental species could be more suitable for comparison with living species, but it might require longer calculations.

6.5.3. *Unique genetic code and the emergence of a multi-species ecosystem*

All living cells use the same code that is responsible for the transcription of information from DNA to proteins [74, 75]. It suggests that at a certain point of evolution of life on Earth a replicator that invented this apparently effective mechanism was able to eliminate replicators of all other species (if they existed) and establish, at least for a short time, a single-species ecosystem. Although this process is still to a large extent mysterious, one expects that subsequent evolution of these successful replicators leads to their differentiation and proliferation of species. In such a way the ecosystem shifted from a single- to multi-species one [76]. It seems to us that our model might provide some insight into this problem. Numerical simulations show [56, 57] that the oscillatory behavior appears in our model only for the relative update rate $r < 0.27$. When preys reproduce faster ($r > 0.27$), a different behavior can be seen and the model reaches a steady state with almost all predators belonging to the same species with the size m close to 1. Only from time to time a new species is created with even larger m and a change of the dominant species might take place. In our opinion, it is possible that at the very early period of evolution of life on Earth, the ecosystem resembled the case $r > 0.27$. This is because at that time substrates ('preys') were renewable faster than primitive replicators ('predators') could use them. If so, every invention of the increase of the efficiency ('size') could invade the entire system. In particular, the invention of the coding mechanism could spread over the entire system. A further evolution increased the efficiency of predators and that effectively shifted the (single-species) ecosystem toward the $r < 0.27$ (multi-species, oscillatory) regime.

6.5.4. *Multispecies prey-predator model - summary and perspectives*

In this section we discussed a model where densities of preys and predators as well as the number of species show long-term oscillations, even though the dynamics of the model is not exposed to any external periodic forcing. It suggests that the oscillatory behavior of the Earth ecosystem predicted by Raup and Sepkoski could be simply a natural feature of its dynamics and not the result of an external factor. Some predictions of our model such as the lifetime of species or the time dependence of their population sizes might be testable against palaeontological data. Certainly, our model is based on some restrictive assumptions that drastically simplify the complexity of the real ecosystem. Nevertheless, it includes some of its important ingredients: replication, mutation, and competition for resources (food and space). As an outcome, the model shows that typically there is no equilibrium-like solution and the ecosystem remains in an evolutionary cycle. The model does not include geographical barriers but let us notice that palaeontological data that suggest the periodicity of mass extinctions are based only on marine fossils [66]. More realistic versions should take into account additional trophic levels, gradual mutations, or sexual reproduction. One should also notice that the palaeontological data are mainly at a genus, and not species level. It would be desirable to check whether the behavior of our model is in some sense generic or it is merely a consequence of its specific assumptions. An interesting possibility in this respect could be to recast our model in terms of Lotka-Volterra like equations and use the methodology of adaptive dynamics developed by Dieckmann *et al.* [68]. Of course, the real ecosystem was and is exposed to a number of external factors such impacts of astronomical objects, volcanism or climate changes. Certainly, they affect the dynamics of an ecosystem and contribute to the stochasticity of fossil data. Filtering out these factors and checking whether the main evolutionary rhythm is indeed set by the ecosystem itself, as suggested in the present paper, is certainly a difficult task but maybe worth an effort.

6.6. Computational approaches to the evolution of language

In this section we describe computational approaches to the problem of evolution of language. In this field the mainstream research takes the darwinian standpoint: natural selection guided the language development and emergence of its basic features. One thus accept that at least some features of language have certain adaptive value and their gradual development is much more plausible than a catastrophic change. There are, however, some issues that still remain unclear within such a darwinian approach. For example evolutionary development of a reliable communication system requires a substantial amount of altruism and it is not clear whether standard explanations that refer to kin selection or reciprocal altruism are applicable (for example kin selection does not explain our willingness to talk to non-kin). Another problem is concerned with the interaction of evolution and learning, sometimes known as a Baldwin effect. In some cases, learning is known to direct the evolutionary changes, and perhaps in such a way humans developed a language-specific adaptations commonly termed Language Acquisition Device (LAD). Efficiency of the Baldwin effect and even the very existence of LAD remain, however, open problems. There is perhaps a little chance that computational modeling will definitely resolve these issues. But already at the level of constructing appropriate models one has to quantify relevant processes and effects, and even that can provide a valuable insight.

6.6.1. *Evolution and language development*

The ability to use language distinguishes humans from all other species. Certain species also developed some communication modes but of much smaller capabilities as well as complexity. Since several decades various schools are trying to explain the emergence and development of language. Natives argue that language capacity is a collection of domain-specific cognitive skills that are somehow encoded in our genome. However, the idea of the existence of such a Language Acquisition Device or "language organ" (the term coined by their most prominent representative Noam Chomsky [77]), was challenged by empiricists, who argue that linguistic performance of humans can be explained using domain-general learning techniques. The recent critique along this line was made by Sampson [78], who questions even the most appealing argument of natives, that refer to the poverty of stimulus and apparently fast learning of grammar by children. An important issue of possible adaptative merits of language does not seem to be settled either. Non-adaptationists, again with Chomsky as the most famous representative [79], consider language as a side effect of other skills and thus claim that its evolution, at least at the beginning, was not related with any fitness advantage. A chief argument against the non-adaptationist stand is the observation that there is a number of costly adaptations that seem to support human linguistic abilities such as a large brain, a longer infancy period or descended larynx. Recently, in their influential paper, Pinker and Bloom argued that, similarly to other complex adaptations, language evolution can only be explained by means of natural selection mechanisms [80]. Their paper triggered a number of works where language was examined from the perspective of evolutionary biology or game theory [81, 82]. In particular, Nowak *et al.* used some optimization arguments, that might explain the origin of some linguistic universals [83, 84]. They suggest that words appeared in order to increase the expressive capacity and sentences (made of words) limit memory requirements. Confrontation of natives with empiricists and adaptationists with non-adaptationists so far does not seem to lead to consensus but certainly deepened our understanding of these problems [85].

Recently, a lot of works on the language emergence seem to have an evolutionary flavor. Such an approach puts some constraints on possible theories of the language origin. In particular, it rules out non-adaptationist theories, where language is a mere by-product of having a large and complex brain [86]. The emergence of language has been also listed as one of the major transitions in the evolution of life on Earth [6]. An interesting question is whether this transition was variation or selection limited [87]. In variation limited transitions the required configuration of genes is highly unlikely and it takes a considerable amount of time for the nature to invent it. For selection limited transitions the required configuration is easy to invent but there is no (or only very weak) evolutionary pressure that would favor it. Relatively large cognitive capacities of primates and their genetic proximity with humans suggests that some other species could have been also capable to develop language-like communication. Since they did not, it was perhaps due to a weak selective pressure. Such indirect arguments suggest that the emergence of language was selection limited [87].

Some interesting results can be obtained by applying game-theory reasoning to one of the most basic problems of emerging linguistic communication, namely why do we talk (at all!) and why do we exchange valuable and trustful information. Since speaking is costly (it takes time, energy and sometimes might expose a speaker to predators), and listening is not, such a situation seems to favor selfish individuals that would only listen but would not speak. Moreover, in the case of the conflict of interests the emerging communication system would be prone to misinformation or lying. The resolution of these dilemmas usually refers to the kin selection [88] or reciprocal altruism [89]. In other words, speakers remain honest because they are helping their relatives or they expect that others will do the same for them in the future. As an alternative explanation Dessalles [90] suggests that honest information is given freely because it is profitable - it is a way of competing for status within a group. Computational modelling of Hurford [91] gives further evidence that speaking might be more profitable than listening. Hurford considered agents engaged in communicative tasks (one speaker and one hearer) and their abilities evolved with the genetic algorithm that was set to prefer either communicative or interpretative success. Only in the former case the emerging language was similar to natural languages were synonymy was rare and homonymy tolerated. When interpretative success was used as the basis of selection then the converse situation (unknown in natural language) arose: homonymy was rare and synonymy tolerated. Some related results on computational modeling of the honest cost-free communication are reported by Noble [92].

A necessary ingredient of language communication is learning. It is thus legitimate to ask whether darwinian selection might be responsible for the genetic hard-wiring of a Language Acquisition Device. Indeed, this (to some extent hypothetical) organ is most likely responsible for some of the arbitrary (as opposed to the functional) linguistic structures. But for such an organ to be of any value, an individual first has to acquire the language. The inheritance of characteristics acquired during an individual lifetime is usually associated with discredited lamarckian mechanism and thus considered to be suspicious. However, the relation between evolution and learning is more delicate and the attempts to clarify the mutual interactions of these two adaptive mechanisms have a long history. According to a purely darwinian explanation, known as a Baldwin effect [13, 93, 94], there might appear a selective pressure in a population for the evolution of the instinctive behavior that would replace the beneficial, but costly, learned behavior [95]. Baldwin effect presumably played an important role in the emergence and evolution of language but certain aspects of these processes still remain unresolved [14]. For example, one of the assumptions that is needed for the Baldwin effect to be effective is a relatively stable environment since otherwise rather slow evolutionary processes will not catch up with the fast changing environment. Since the language formation processes are rather fast (in comparison to the evolutionary time scale), Christiansen and Chater questioned the role of adaptive evolutionary processes in the formation of arbitrary structures like Language Acquisition Device [96]. Actually, they suggest a much different scenario, where it is a language that adapted to human brain structures rather than vice versa.

6.6.2. *Language as a complex adapting system*

From the above description it is clear that studying of the emergence and evolution of language is a complex and multidisciplinary task and requires cooperation of not only linguists, neuroscientists, and anthropologists, but also experts in artificial intelligence, computer sciences or evolutionary biology [97]. One can distinguish two levels at which language can be studied and described [98] (Fig. 6.6). At the individual level the description centers on the individual language users: their linguistic performance, language acquisition, speech errors, speech pathologies or brain functioning in relation with language processing. At the individual level the language of each individual is slightly different. Nevertheless, within certain population these individuals can efficiently communicate and that establishes the population level. At this level the language is considered as an abstract system that exists in a sense separately from the individuals users. There are numerous interactions between these two levels. Indeed, the linguistic behavior of individuals depends on the language (at the population level) specific to the population they are part of. And, as a feedback, the language used in a given population is a collective behavior and emerges from linguistic behavior of individuals composing this population. Various processes shaping such a complex system are operating at different time scales. The fastest dynamics is operating at the individual level (ontogenetic timescale [99]) that includes, for example, language acquisition processes. Much slower processes, such as migrations of language populations, dialects formation or language extinctions, are operating at the so-called glossogenetic timescale. The slowest processes govern the biological evolution of language users and that defines the phylogenetic timescale. Processes operating at these different timescales are not independent (Fig. 6.6). Biological evolution might change linguistic performance of individuals and that might affect the glossogenetic processes. For example, a mutation that changes the vocal ability of a certain individual, if spread in his/her population, might lead to a dialect formation or a language extinction. Such population-level processes might change the selective pressure that individual language users are exposed to and that might affect phylogenetic processes, closing thus the interaction loop.

Various levels of descriptions and processes operating at several timescales suggest that complex models must be used to describe adequately the language evolution. Correspondingly, the analysis of such models and predicting their behavior also seem to be difficult. It is known that some phenomena containing feedback interactions might be described in terms of nonlinear differential equations, such as for example already described Lotka-Volterra equations. The behavior of such nonlinear equations is often difficult to predict, since abrupt changes even of the qualitative nature of solutions might take place. Language evolution is, however, much more complex than ecological problems of interacting populations and its description in terms of differential equations would be much more complicated if at all feasible. It seems that recently the most promising and frequently used approach to examine such systems is computational modelling of multi-agent systems. Using this method one examines a language that emerges in a bottom-up fashion as a result of interactions within a group of agents equipped with some linguistic functions. Then one considers language as a complex adaptive system that evolves and complexities according to biologically inspired principles such as selection and self-organization [100]. Thus, the emerging language is not static but evolves in a way that hopefully is similar to human language evolution. Of course, using such an approach one cannot explain all intricacies of human languages. A more modest goal would be to understand some rather basic features that are common to all languages such as meaning-form mappings, origin of linguistic coherence (among agents without central control and global view), or coevolutionary origin of grammar and meaning.

Within such a multi-agent approach, two groups of models can be dis-

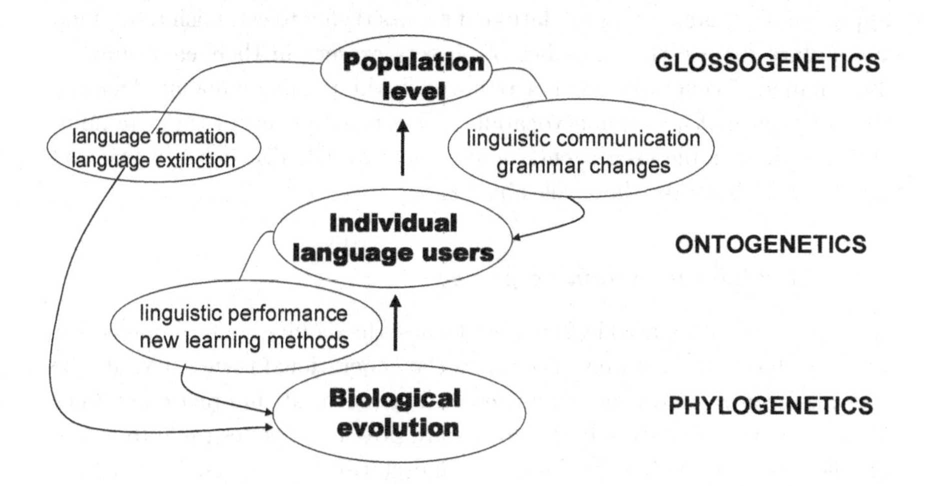

Fig. 6.6. Language as a complex adaptive system. Many different processes governing the language evolution are entangled at various levels. Relatively fast individual level (ontogenetics), comprising *e.g.*, language acquisition processes, is determined mainly by interactions between individual language users. Much slower are populational-level processes (glossogenetics) such as language formations, extinctions, grammar changes or migrations. To obtain a complete description one has to consider also biological evolution (phylogenetics) and these are the slowest processes of the language evolution. Various processes at individual and population level affect the fitness landscape and that influences the biological evolution level. Similarly, individual language user level is affected by populational level processes.

tinguished. In the first one, originating from the so-called iterated learning model, one is mainly concerned with the transmission of language between successive generations of agents [101, 102]. Agents that are classified as teachers produce some expressions that are passed to learners that try to infer their meaning using statistical learning techniques such as neural networks. After a certain number of iterations teachers are replaced by learners and a new population of learners is introduced. The important issue that the iterated learning model has successfully addressed is the transition from holistic (complex meaning expressed by a single form) to compositional language (composite meaning is expressed with composite form). However, since such a procedure is computationally relatively demanding and the number of communicating agents is thus typically very small, the problem of the emergence of linguistic coherence must be neglected in this approach.

To tackle this problem Steels introduced a naming game model [11]. In this approach one examines a population of agents trying to establish a common vocabulary for a certain number of objects present in their environment. The change of generations is not required in the naming game model since the emergence of a common vocabulary is a consequence of the communication processes between agents, and agents are not divided into teachers and learners but take these roles in turn.

6.6.3. *Evolutionary naming game*

It seems that the iterated learning model and the naming-game model are at two extremes: the first one emphasizes the generational turnover while the latter concentrates on the single-generation (cultural) interactions. Since in the language evolution both aspects are present, it is desirable to examine models that combine evolutionary and cultural processes. Recently we have introduced such a model [12, 103] and below we briefly describe its properties.

In our model we consider a set of agents located at sites of the square lattice of the linear size N. Agents are trying to establish a common vocabulary on a single object present in their environment. An assumption that agents communicate only on a single object does not seem to restrict the generality of our considerations and has already been used in some other studies of naming game [104, 105] or language-change [106, 107] models. A randomly selected agent takes the role of a speaker that communicates a word chosen from its inventory to a hearer that is randomly selected among nearest neighbors of the speaker. The hearer tries to recognize the communicated word, namely it checks whether it has the word in its inventory. A positive or negative result translates into communicative success or failure, respectively. In some versions of the naming game model [104, 105] a success means that both agents retain in their inventories only the chosen word, while in the case of failure the hearer adds the communicated word to its inventory.

To implement the learning ability we have modified this rule and assigned weights w_i ($w_i > 0$) to each i-th word in the inventory. The speaker selects then the i-th word with the probability $w_i / \sum_j w_j$ where summation is over all words in its inventory (if its inventory is empty, it creates a word randomly). If the hearer has the word in its inventory, it is recognized. In addition, each agent k is characterized by its learning ability l_k ($0 < l_k < 1$), that is used to modify weights. Namely, in the case of success both speaker and hearer increase the weights of the communicated word by their learning abilities, respectively. In the case of failure the speaker subtracts its learning ability from the weight of the communicated word. If after such a subtraction a weight becomes negative, the corresponding word is removed from the repository. The hearer in the case of failure, *i.e.*, when it does not have the word in its inventory, adds the communicated word to its inventory with a unit weight.

Apart from communication, agents in our model evolve according to the population dynamics: they can reproduce, mutate, and eventually die. To specify intensity of these processes we have introduced the communication probability p. With the probability p the chosen agent becomes a speaker and with the probability $1-p$ a population update is attempted. During such a move the agent dies with the probability $1-p_{\rm surv}$, where $p_{\rm surv} = \exp(-at)[1-\exp(-b\sum_j w_j/\langle w\rangle)]$, and $a \sim 0.05$ and $b = 5$ are certain parameters whose role is to ensure a certain speed of population turnover. Moreover, t is the age of an agent and $\langle w\rangle$ is the average (over agents) sum of weights. Such a formula takes into account both its linguistic performance (the bigger $\sum_j w_j$ the larger $p_{\rm surv}$) and its age. If the agent survives (it happens with the probability $p_{\rm surv}$), it breeds, provided that there is an empty site among its neighboring sites. The offspring typically inherits parent's learning ability and the word from its inventory that has the highest weight. In the offspring's inventory the weight assigned initially to this word equals one. With the small probability $p_{\rm mut}$ a mutation takes place and the learning ability of an offspring is selected randomly anew. With the same probability an independent check is made whether to mutate the inherited word. Numerical simulations show that the described below behavior of our model is to some extent robust with respect to some modifications of its rules. For example, qualitatively the same behavior is observed for modified parameters a and b, different form of the survival probability $p_{\rm surv}$ (provided it is a decreasing function of t and an increasing function of $\sum_j w_j$), or different breeding and/or mutation rules. To examine the behavior of the model we have measured the communication success rate s defined as an average over agents and simulation time of the fraction of successes with respect to all communication attempts. Moreover, we have measured the average learning ability l.

Our model captures all three basic aspects of language: learning, culture, and evolution. Agents in our model are equipped with an evolutionary trait: learning ability. When communication between agents is sufficiently frequent (*i.e.*, when p is large enough), cultural processes create a niche in

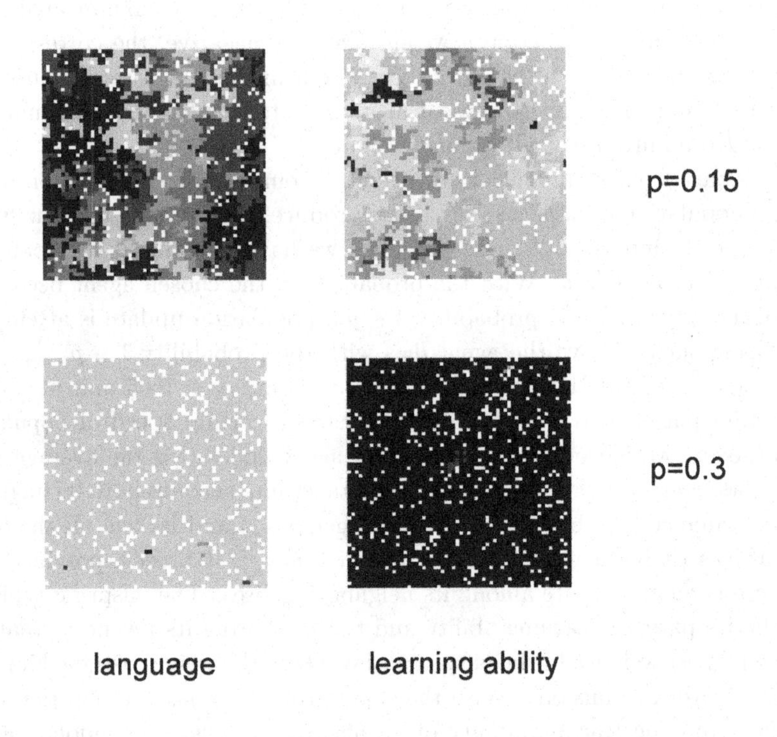

Fig. 6.7. Exemplary configurations of the evolutionary naming game model [12, 103] with $L = 60$ and $p_{\rm mut} = 0.001$. In the small-p phase (upper panel) communications are infrequent and agents using the same language (left)or having the same learning abilities (right) form only small clusters. In this phase the communication success rate s and the learning ability l are small (see also Fig. 6.8). The larger the learning ability of an agent the darker are pixels representing it (white: l=0; black: l=1). In the large-p phase (lower panel) frequent communications result in the emergence of the common language. Moreover, almost all agents use the same language and have the same, and large, learning ability.

which a larger learning ability becomes advantageous. It causes an increase of learning ability, but its large value in turn makes the cultural processes more efficient. As a result the model was shown to undergo an abrupt bio-linguistic transition where both linguistic performance (s) and ability (l) of agents change very rapidly (see Figs. 6.7-6.8) [12]. It was also shown

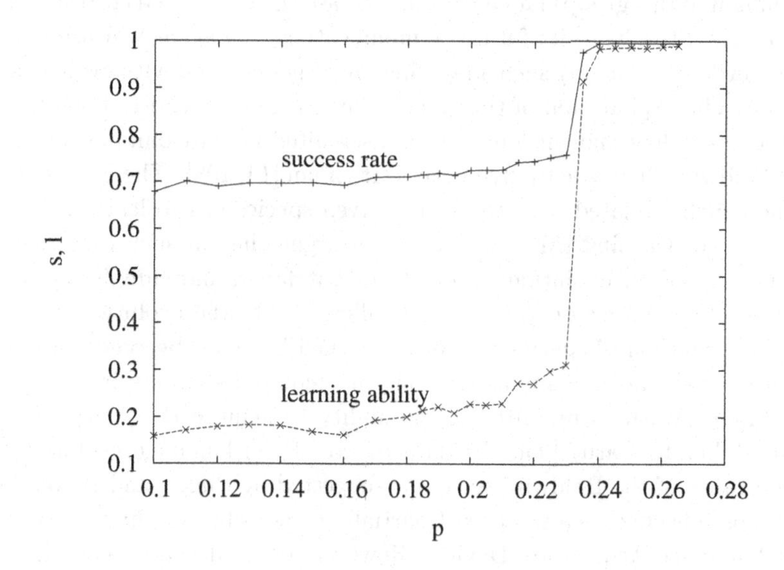

Fig. 6.8. The success rate s and the learning ability l as a function of the communication probability p. Calculations were made for system size $L = 60$ and mutation probability $p_{\mathrm{mut}} = 0.001$. Simulation time for each value of p was typically equal to 10^5 steps with $3 \cdot 10^4$ steps discarded for relaxation. A step is defined as a single, on average, update of each site.

that under the plausible assumption, that the intensity of communication increases continuously in time, this bio-linguistic transition is replaced with a series of fast, transition-like changes [103]. In our opinion, the proposed model shows that linguistic and biological processes have a strong influence on each other, which has certainly contributed to an explosive development of our species.

6.6.4. *Baldwin effect*

That learning in our model modifies the fitness landscape of a given agent and facilitates the genetic accommodation of learning ability is actually a manifestation of the much debated Baldwin effect. The fact that the success rate s and the learning ability l have a jump at the same value of p shows that communicative and biological ingredients in our model strongly influence each other and that leads to the single and abrupt transition. In our model successful communication requires learning. A new-born agent communicating with some mature agents who already worked out a certain (common in this group) language will increase the weight of a corresponding word. As a result, in its future communications the agent will use mainly this word. In what way such a learning might get coupled with evolutionary traits? The explanation of this phenomenon is known as a Baldwin effect. Although at first sight it looks like a discredited Lamarckian phenomenon, the Baldwin effect is actually purely Darwinian [14, 108]. There are usually some benefits related with the task a given species has to learn and there is a cost of learning this task. One can argue that in such case there is some kind of an evolutionary pressure that favors individuals for which the benefit is larger or the cost is smaller. Then, the evolution will lead to the formation of species where the learned behavior becomes an innate ability. It should be emphasized that the acquired characteristics are not inherited. What is inherited is the ability to acquire the characteristics (the ability to learn) [95]. In the context of the language evolution the importance of the Baldwin effect was suggested by Pinker and Bloom [80]. Perhaps this effect is also at least partially responsible for the formation of the Language Acquisition Device. However, many details concerning the role of the Baldwin effect in the evolution of language remain unclear [109].

We already argued [12], that in our model the Baldwin effect is also at work. Let us consider a population of agents with the communication probability p below the threshold value ($p = p_c \approx 0.23$). In such a case the learning ability remains at a rather low level (since clusters of agents using the same language are small, it does not pay off to be good at learning the language of your neighbors). Now, let us increase the value of p above the threshold value. More frequent communication changes the behavior dramatically. Apparently, clusters of agents using the same language are now sufficiently large and it pays off to have a large learning ability because that increases the success rate and thus the survival probability $p_{\rm surv}$. Let us notice that $p_{\rm surv}$ of an agent depends on its linguistic performance ($\sum_j w_j$) rather than its learning ability. Thus clusters of agents of good linguistic performance (learned behavior) can be considered as niches that direct the evolution by favoring agents with large learning abilities, which is precisely the Baldwin effect. It should be noticed that linguistic interactions between agents (whose rate is set by the probability p) are typically much faster than evolutionary changes (set by $p_{\rm mut}$) and such an effect was also observed in simulations [12].

As a result of a positive feedback (large learning ability enhances communication that enlarges clusters that favors even more the increased learning ability) a discontinuous transition takes place both with respect to the success rate and learning ability . An interesting question is whether such a behavior is of any relevance in the context of human evolution. It is obvious that development of language, which probably took place somewhere around 10^5 years ago, was accompanied by important anatomical changes such as fixation of the so-called speech gene (FOXP2), descended larynx or enlargement of brain [110]. Linguistic and other cultural interactions that were already emerging in early hominid populations were certainly shaping the fitness landscape and that could direct the evolution of our ancestors via the Baldwin effect.

The examined model is not very demanding computationally. It seems to be possible to consider agents talking on more than one object [111], or to examine statistical properties of simulated languages such as for example, distributions of their lifetimes or of the number of users. It would be interesting to examine the role of topology of interaction network and place agents on complex networks, like *e.g.*, scale-free networks, that are known to provide a more realistic description of human linguistic interactions [105]. One can also study diffusion of languages, the role of geographical barriers [112], or formation of language families. There is already an extensive literature documenting linguistic data as well as various computational approaches modeling, for example, competition between already existing natural languages [113–115]. The dynamics of the present model, that is based on an act of elementary communication, offers perhaps more natural description of dynamics of languages than some other approaches that often use some kind of coarse-grained dynamics.

6.7. Conclusions

In the present paper we reviewed computational methods that are used for modeling evolutionary systems. We emphasized the need and advantages of using models with individual-based dynamics. We also drew attention to various time scales of processes that shape the evolution of complex systems. In ecosystems these are ecological and evolutionary time scales. In the language evolution cultural processes set an additional timescale. Perhaps the most interesting phenomena arise from interactions of processes of various times scales. Evolutionary cycling or Baldwin effect are excellent examples of such phenomena, to claim however their satisfactory understanding, much remains to be done.

This mini-review is of course biased by our own experience in this field. We did not even mention about a number of other approaches and techniques of modeling evolutionary systems. Some of them are covered in other chapters of this volume.

Acknowledgments

The present paper is based on the lecture that A.L. delivered during the conference "From Genetics to Mathematics" (Zbąszyń, Poland, October-2007). A.L. thanks the organizers of the conference for invitation. Our project is supported with the research money allocated for the period 2008-2010 under the grant N N202 071435. We also acknowledge access to the computing facilities at Poznań Supercomputer and Networking Center.

References

[1] N. Eldredge and S. J. Gould, Punctuated Equilibria: An Alternative to Phyletic Gradualism. In eds. T. J. M. Schopf, *Models in Palaeobiology*, pp. 82-115. Freeman, Cooper, San Francisco, (1972).

[2] G. G. Simpson, *Fossils and the history of life.* (Scientific American Library, New York, 1983).

[3] M. E. J. Newman and R. G. Palmer, *Modelling Extinction.* (Oxford University Press, New York, 2003).

[4] P. Bak and K. Sneppen, Punctuated equilibrium and criticality in a simple model of evolution, *Phys. Rev. Lett.* **71**, 4083–4086, (1993).

[5] D. M. Raup and J. J. Sepkoski, Periodicities of extinctions in the geologic past, *Proc. Natl. Acad. Sci. U.S.A.* **81** 801–805, (1984).

[6] J. Maynard-Smith and E. Szathmáry, *The Major Transitions in Evolution.* (Oxford University Press, New York, 1997).

[7] R. Dawkins, *The Selfish Gene.* (Oxford University Press, 1976).

[8] M. H. Christiansen and S. Kirby, *Language Evolution: The Hardest Problem in Science.* In eds. M. H. Christiansenand S. Kirby, *Language Evolution.* (Oxford University Press Inc., New York, 2003).

[9] G. F. Fussmann, S. P. Ellner and N. G. Hairston Jr., Evolution as a critical component of plankton dynamics, *Proc. Roy. Soc. B* **270**, 1015–1022, (2003).

[10] J. N. Thompson, Rapid evolution as an ecological process, *Trends in Ecol. Evol.* **13**, 329–332, (1998).

[11] L. Steels, A self-organizing spatial vocabulary, *Artif. Life* **2**, 319–332, (1995).

[12] A. Lipowski and D. Lipowska, Bio-linguistic transition and the Baldwin effect in the evolutionary naming game model, *Int. J. Mod. Phys. C* **19**, 399–407, (2008).

[13] J. M. Baldwin, A new factor in evolution, *American Naturalist* **30**, 441-451, (1896).

[14] H. Yamauchi, Baldwinian Accounts of language evolution, Ph.D. thesis, University of Edinburgh, Edinburgh, Scotland (2004).

[15] P. F. Verhulst, Notice sur la loi que la population poursuit dans son accroissement, *Corresp. Math. Phys.* **10**, 113–121, (1838).

[16] A. J. Lotka, Analytical note on certain rhythmic relations in organic systems, *Proc. Natl. Acad. Sci. USA* **6**, 410–415, (1920).

[17] V. Volterra, Variazioni e fluttuazioni del numero dŠindividui in specie animali conviventi, *Mem. Accad. Nazionale Lincei*, ser. 6, **2**, 31-112, (1926).

[18] J. D. Murray, *Mathematical Biology Vols. I/II.* (Springer-Verlag, New York, 2002).

[19] R. May, Limit cycles in predator-prey communities, *Science* **177**, 900-902, (1972).

[20] E. E. Holmes, M. A. Lewis, J. Banks, and R. R. Veit, Partial differential equations in ecology: spatial interactions and populations dynamics, *Ecology* **75**, 17-29, (1994).

[21] M. P. Hassell, H. N. Comins, and R. M. May, Spatial structure and chaos in insect population dynamics, *Nature, Lond.* **353**, 255-258, (1991).

[22] D. P. Hardin, P. Takac, and G. F. Webb, Dispersion population models discrete in time and continuous in space, *J. Math. Biol.* **28** 1–20, (1990).

[23] A. Lipowski, Oscillatory behaviour in a lattice prey-predator system, *Phys. Rev. E* **60**, 5179–5184, (1999).

[24] A. Lipowski and D. Lipowska, Nonequilibrium phase transition in a prey-predator system, *Physica A* **276**, 456–464, (2000).

[25] M. Kowalik, A. Lipowski and A. L. Ferreira, Oscillations and dynamics in a two-dimensional prey-predator system, *Phys. Rev. E* **66**, 066107-1–066107-5, (2002).

[26] H. Hinrichsen, Nonequilibrium Critical Phenomena and Phase Transitions into Absorbing States, *Adv.Phys.* **49** 815-958, (2000).

[27] A. Lipowski and M. Droz, Phase transitions in nonequilibrium d-dimensional models with q absorbing states, *Phys. Rev. E* **65**,056114-1 –056114-7, (2002).

[28] P. Grassberger,and A. de la Torre, Reggeon field theory (SchlöglŠs first model) on a lattice; Monte-Carlo calculations of critical behaviour, *Ann. Phys. (N.Y.)* **122**, 373Ű-396, (1979).

[29] G. Grinstein, D. Mukamel, R. Seidin, and Ch. H. Bennett, Temporally periodic phases and kinetic roughening. *Phys. Rev. Lett.* **70**, 3607–3610 (1993).

[30] M. Mobilia, I. T. Georgiev, and U. C. Taüber, Fluctuations and correlations in lattice models for predator-prey interaction. *Phys. Rev. E* **73**, 040903, (2005).

[31] M. Kowalik, Badanie oscylacji czasowych w układach makroskopowych. Ph.D. thesis, Adam Mickiewicz University, Poznań Poland (2003). (in Polish)

[32] R. Monetti, A. F. Rozenfeld and E. F. Albano, Study of interacting particle systems: the transition to the oscillatory behavior of a prey-predator model. *Physica A* **283**, 52–58, (2000).

[33] A. F. Rozenfeld and E. V. Albano, Critical and oscillatory behavior of a system of smart preys and predators. *Phys. Rev. E* **63**, 061907, (2001).
[34] R. Rai and H. Singh, Stochastic resonance without an external periodic drive in a simple prey-predator model. *Phys. Rev. E* **62**, 8804–8807, (2000).
[35] H. Gang, T. Ditzinger,C. Z. Ning and H. Haken, Stochastic resonance without external periodic force. *Phys. Rev. Lett.* **71**, 807–810, (1993).
[36] D. Stauffer, Percolation clusters as teaching aid for Monte-Carlo simulation and critical exponents. *Am. J. Phys*, **45**, 1001, (1977).
[37] D. Stauffer, Scaling theory of percolation clusters. *Phys. Rep.* **54**, 1, (1979).
[38] T. Antal and M. Droz, Phase transitions and oscillations in a lattice prey-predator model. *Phys. Rev. E* **63**, 056119, (2001).
[39] N. Boccara, O. Roblin, and M. Roger, Automata network predator-prey model with pursuit and evasion. *Phys. Rev. E* **50**, 4531–4541, (1994).
[40] H. Matsuda, N. Ogita, A. Sasaki, and K. Sato, Statistical mechanics of population: the Lotka-Volterra model. *Prog. Theor. Phys.* **88**, 1035-1049, (1992).
[41] A. Pękalski, A short guide to predator-prey lattice models. *Computing in Science and Engineering* **6**, 62–66, (2004).
[42] A. Provata, G. Nicolis and F. Baras, Ocillatory dynamics in low dimensional lattices: a lattice Lotka-Volterra Model. *J. Chem. Phys.* **110**, 8361–8368, (1999).
[43] J. E. Satulovsky and T. Tomé, Stochastic lattice gas model for a predator-prey system, *Phys. Rev. E* **49**, 5073–5079, (1994).
[44] B. Blasius, A. Huppert, and L. Stone, Complex dynamics and phase synchronization in spatially extended ecological systems, *Nature* **399** 354–359, (1999).
[45] S. Engen and B. E. Sæther, Generalizations of the Moran Effect Explaining Spatial Synchrony in Population Fluctuations, *Amer. Natur.* **166**, 603–612, (2005).
[46] W. Schaaffer, Stretching and folding in lynx fur returns: Evidence for a strange attractor in nature, *Am. Nat.* **124**, 798-820, (1984).
[47] P. D. Gingerich, (1983). Rates of evolution: effects of time and temporal scaling, *Science* **222**, 159Ũ-161, (1983).
[48] A. I. Khibnik and A. S. Kondrashov, Three mechanisms of the Red Queen dynamics, *Proc. R. Soc. Lond. B* **264**, 1049-1056, (1997).
[49] N. Vandevalle and M. Ausloos, The robustness of self-organized criticality against extinctions in a tree-like model of evolution, *Europhys. Lett.* **32**, 613-618, (1995).
[50] R. V. Solé and S. C. Manrubia, Extinction and self-organized criticality in a model of large-scale evolution, *Phys. Rev. E* **54**, R42-R45, (1996).
[51] L. A. N. Amaral and M. Meyer, Environmental changes, coextinction, and patterns in the fossil record, *Phys. Rev. Lett.* **82**, 652-655, (1999).
[52] G. Abramson, Ecological model of extinctions, *Phys. Rev. E* **55**, 785-788, (1997).
[53] G. Caldarelli, P. G. Higgs, and A. J. McKane, Modelling coevolution in multispecies communities, *J. Theor. Biol.* **193**, 345-358, (1998).

[54] F. Coppex, M. Droz, and A. Lipowski, Extinction dynamics of Lotka-Volterra ecosystems on evolving networks, *Phys. Rev. E* **69**, 061901, (2004).
[55] D. Chowdhury, D. Stauffer, and A. Kunwar, Uniffication of small and large time scales for biological evolution: deviations from Power Law, *Phys. Rev. Lett.* **90**, 068101, (2003).
[56] A. Lipowski, Periodicity of mass extinctions without an extraterrestrial cause, *Phys. Rev. E* **71**, 052902–052905, (2005).
[57] A. Lipowski and D. Lipowska, Long-term evolution of an ecosystem with spontaneous periodicity of mass extinctions, *Theory in Biosciences* **125**, 67–77, (2006).
[58] C. Patterson and A. B. Smith, Periodicity in extinction: the role of the systematics, *Ecology* **70**, 802–811, (1989).
[59] W. T. Fox, Harmonic analysis of periodic extinctions, *Paleobiology* **13**, 257–271, (1987).
[60] R. E. Plotnick and J. J. Sepkoski, A multiplicative multifractal model for originations and extinctions, *Paleobiology* **27**, 126–139, (2001).
[61] A. Prokoph, A. D. Fowler, and R. T. Patterson, Evidence for periodicity and nonlinearity in a high-resolution fossil record of long-term evolution, *Geology* **28**, 867–870, (2000).
[62] M. Davis, P. Hut, and R. M. Muller, Extinction of species by periodic comet showers. *Nature* **308**, 715–717, (1984).
[63] M. R. Rampino and R. B. Stothers, Terrestrial mass extinctions, cometary impacts and the Sun's motion perpendicular to the galactic plane, *Nature* **308**, 709–712, (1984).
[64] R. B. Stothers, Flood basalts and extinction events, *Geophys. Res. Lett.* **20**, 1399–1402, (1993).
[65] R. B. Stothers, Periodicity of the Earth's magnetic reversals, *Nature* **322**, 444–446, (1986).
[66] R. A. Rohde and R. A. Muller, Cycles in fossil diversity, *Nature* **434**, 208–210, (2005).
[67] A. I. Miller and M. Foote, Increased longevities of post-maleozoic marine genera after mass extinctions, *Science* **302**, 1030–1032, (2003).
[68] U. Dieckmann, P. Marrow, and R. Law, Evolutionary cycling in predator-prey interactions: population dynamics and the red queen, *J. Theor. Biol.* **176**, 91–102, (1995).
[69] S. M. Stanley, Delayed recovery and the spacing of major extinctions, *Paleobiology* **16**, 401–414, (1990).
[70] R. E. Plotnick and M. L. McKinney, Ecosystem organization and extinction dynamics, *Palaios* **8**, 202–212, (1993).
[71] S. M. Stanley, An explanation for Cope's rule, *Evolution* **27**, 1–26, (1973).
[72] J. Alroy, Cope's rule and the dynamics of body mass evolution in north American Fossil mammals, *Science* **280**, 731–734, (1998).
[73] B. Van Valkenburgh, X. Wang, and J. Damuth, Cope's rule, hypercarnivory, and extinction in north American Canids, *Science* **306**, 101–104, (2004).
[74] L. E. Orgel, Molecular replication, *Nature* **358**, 203–209, (1992).
[75] E. Szathmáry, The origin of the genetic code, *Trends in Genetics* **15**, 223–

229, (1999).

[76] A. Lipowski, Multiplicity of species in some replicative systems, *Phys. Rev. E* **61**, 3009–3014 (2000).

[77] N. Chomsky, *Aspects of the theory of syntax.* (MIT Press, Cambridge, MA, 1965).

[78] G. Sampson,*Educating Eve: The 'Language Instinct' Debate.* (Cassell, London, 1999).

[79] N. Chomsky, *Language and Mind.* (Harcourt Brace Jovanovich, San Diego, 1972).

[80] S. Pinker and P. Bloom, Natural language and natural selection, *Behav. Brain Sci.* **13**, 707–784, (1990).

[81] R. S. Jackendoff, *Languages of the mind.* (MIT Press, 1992).

[82] C. Knight, M. Studdert-Kennedy, and J. Hurford, Eds. *The Evolutionary Emergence of Language Social Function and the Origin of Linguistic Form.* (Cambridge University Press, 2000).

[83] M. A. Nowak and N. L. Komarova, Towards an evolutionary theory of language, *Trends in Cogn. Sci.* **5**, 288–295, (2001).

[84] M. A. Nowak and D. C. Krakauer, The evolution of language. *Proc. Natl. Acad. Sci. USA* **96**, 8028–8033, (1999).

[85] K. Smith, The Transmission of Language: models of biological and cultural evolution. Ph.D. thesis, The University of Edinburgh, Scotland, (2003).

[86] S. J. Gould and R. C. Lewontin, The spandrels of san marco and the panglossian paradigm: a critique of the adaptationist programme, *Proc. Roy. Soc. London B* **205**, 581-598, (1979).

[87] S. Számadó and E. Szathmáry, Language evolution: competing selective scenarios, *Trends. Ecol. Evol.* **21**, 555-561, (2006).

[88] W. D. Hamilton, The genetical theory of social behaviour (I and II), *J. Theor. Biol.* **7**, 1–16, (1964).

[89] R. L. Trivers, The evolution of reciprocal altruism, *Quart. Rev. Biol.* **46**, 35–57, (1971).

[90] J. L. Dessalles, *Altruism, status, and the origin of relevance,* In eds. J. R. Hurford, M. Studdert-Kennedy, and C. Knight, *Approaches to the Evolution of Language: Social and Cognitive Bases* (Cambridge University Press, Cambridge, 1998).

[91] J. R. Hurford, *Why synonymy is rare: Fitness is in the speaker,* In eds. W. Banzhaf, T. Christaller, P. Dittrich, J. T. Kim, and J. Ziegler, *Advances in artificial life - Proceedings of the 7th European Conference on Artificial Life (ECAL), lecture notes in artificial intelligence,* vol. **2801**, pp. 442-451. Springer Verlag, Berlin, (2003).

[92] J. Noble, *Co-operation, competition and the evolution of pre-linguistic communication,* In eds. C. Knight, J. R. Hurford, and M. Studdert-Kennedy, *The Evolutionary Emergence of Language: Social Function and the Origin of Linguistic Form* (Cambridge University Press, Cambridge, 2000).

[93] G. G. Simpson, The Baldwin Effect, *Evolution* **7**, 110–117, (1953).

[94] B. H. Weber and D. J. Depew, Eds., *Evolution and Learning - The Baldwin Effect Reconsidered,* (MIT Press, Cambridge, MA, 2003).

[95] P. Turney, *Myths and legends of the Baldwin Effect*, In eds. T. Fogarty, and G. Venturini, Proc. of the ICML-96, 13th International Conference on Machine Learning, (Bari, Italy, 1996).
[96] M. H. Christiansen and N. Chater, Language as shaped by the brain, Behav. Brain Sci. **31**, 489-509, (2008).
[97] M. A. Nowak, N. L. Komarova, and P. Niyogi, Computational and evolutionary aspects of language, *Nature* **417**, 611–617, (2002).
[98] B. de Boer, *Computer modelling as a tool for understanding language evolution.* In eds. N. Gonthier, J. P. van Bendegem, and D. Aerts, *Evolutionary Epistemology, Language and Culture - A nonadaptationist system theoretical approach*, Dordrecht: Springer, (2006).
[99] S. Kirby, Natural language from artificial life, *Artif. Life* **8**, 185–215, (2002).
[100] L. Steels, *Iterated Learning versus Language Games. Two models for cultural language evolution*, In eds. C. Hemelrijk and E. Bonabeau, *Proceedings of the International Workshop of the Self-Organization and Evolution of Social Behaviour*, University of Zurich, Switzerland, (2002).
[101] H. Brighton, Compositional syntax from cultural transmission, *Artif. Life* **8**, 25–54, (2002).
[102] S. Kirby and J. Hurford, *The emergence of Linguistic Structure; An Overview of the Iterated Learning Model*, In eds. A. Cangelosi and D. Parisi, *Simulating the Evolution of Language*, Springer-Verlag, Berlin, (2001).
[103] A. Lipowski and D. Lipowska, *Computational approach to the emergence and evolution of language - evolutionary naming game model.* E-print: arXiv:0801.1658, (2008).
[104] A. Baronchelli, M. Felici, V. Loreto, E. Caglioti, and L. Steels, Sharp transition towards shared vocabularies in multi-agent systems. *J. Stat. Mech.* **P06014**, (2006).
[105] L. Dall'Asta, A. Baronchelli, A. Barrat, and V. Loreto, Nonequilibrium dynamics of language games on complex networks. *Phys. Rev. E* **74**, 036105, (2006).
[106] D. Nettle, Using Social Impact Theory to simulate language change, *Lingua* **108**, 95–117, (1999).
[107] D. Nettle, Is the rate of linguistic change constant?, *Lingua* **108**, 119–136, (1999).
[108] G. Hinton and S. Nowlan, How learning can guide evolution, *Complex Systems* **1**, 495-502, (1987).
[109] S. Munroe and A. Cangelosi, Learning and the evolution of language: the role of cultural variation and learning cost in the Baldwin Effect, *Artif. Life* **8**, 311-339, (2002).
[110] C. Holden, The origin of Speech, *Science* **303**, 1316-1319, (2004).
[111] A. Lipowski and D. Lipowska, Homonyms and synonyms in the n–object naming game model, E-print: arXiv:0810.3442 (2008).
[112] M. Patriarca and E. Heinsalu, Influence of geography on language evolution, *Physica A* **388**, 296-299, (2008).
[113] D. Abrams and S. H. Strogatz, Modelling the dynamics of language death. *Nature* **424**, 900, (2003).

[114] P. M. C. de Oliveira, D. Stauffer, S. Wichmann, S., and S. M. de Oliveira. A computer simulation of language families, *J. Ling.* **44**, 659–675, (2007).
[115] C. Schulze, D. Stauffer, and S. Wichmann, Birth, survival and death of languages by Monte-Carlo simulation. Commun. Comp. Phys. **3**, 271-294, (2008).

Biological Glossary

Allele - one of two or more alternative forms of a gene, located at corresponding places (loci) on chromosomes.

Allopatric speciation - differentiation of subpopulations isolated by environmental, geographical or biological barriers into two or more new species.

Autosomes - chromosomes which in diploid organism form corresponding (homologous) pairs (chromosomes other than sex chromosomes).

Chargaff's rules: 1. The number of adenine in a DNA molecule equals the number of thymine, the number of guanine equals the number of cytosine. 2. The ratio [G+C]/[A+T] is constant for a species. Rule number 1 is equivalent to the parity rule I and it is a consequence of complementarity of nucleotides in the double helix of DNA.

Chromosome - single, autonomously replicating DNA molecule indispensable for living organisms. Usually, highly organised structure with many protein molecules. In prokaryotic organisms chromosomes usually are circular and located in cytoplasm, in eukaryotic organisms they are linear and located in nucleus.

Codon - tri-nucleotide sequence translated for one amino acid during the protein synthesis or signalling "stop translation".

Codon usage - specific frequencies of using the codons in a given genome, often presented as ratios between frequencies of using synonymous codons.

Complementation - restoring the function of a defective gene by a wild (functional) allele.

Complementing haplotypes - strings of genes in one haplotype which complement the defective alleles at the corresponding fragments of the other haplotype.

Crossover - the exchange of corresponding chromatid (chromosome) segments during meiosis; meiotic recombination.

Diploid - cell or organism having two corresponding sets of chromosomes, usually one set from mother and the other one from father.

Divergence - growing differences between the (genomic) sequences of the same ancestral origin due to accumulation of mutations.

DNA asymmetry - a state of DNA molecule or its fragment which does not fulfil the parity rule II.

Dominant (allele) - an allele, function of which manifest in the heterozygous state.

Gamete - a reproductive cell (usually haploid) which after fusing with the reproductive cell of another sex forms a zygote (diploid).

Gene - there is no good definition of gene.

Genetic code - a set of 64 codons of which 61 are translated for 20 amino acids and three for stop translation signals.

Genetic code degeneracy - character of the genetic code - one amino acid can be coded by more than one codon.

Genetic pool - all genes or genetic information of an interbreeding population (species).

Genome - the whole genetic information of an organism. In diploid organisms, genome is sometimes referred to single complete set of chromosomes - haplotype.

Genotype - corresponds to genome, sometimes used as a genetic content of only one or a few loci.

Haploid - cell or organism having only one set of chromosomes.

Hemizygous - corresponds to genetic information present only in one copy of a gene (*i.e.* men are hemizygous in most of loci of X chromosome).

Heterogametic - individuals of heterogametic sex produce gametes with different sex chromosomes (*i.e.* males of mammals and females of birds).

Heterozygote - an organism having two different alleles in the corresponding locus of homologous chromosomes.

Homogametic - individuals of homogametic sex produce gametes with the same sex chromosomes (*i.e.* females of mammals or males of birds).

Homozygote - an organism having two identical alleles in the corresponding locus of homologous chromosomes.

Inbreeding - mating between relatives.

Inbreeding coefficient - the measure of genetic relations between two individuals, the probability that two genetic elements are inherited from the same ancestor.

Inbreeding depression - reduction of the reproduction potential due to inbreeding.

Inversion - change of the direction of a DNA fragment for 180^0.

Lagging strand - a strand of the DNA molecule which has been synthesized discontinuously, from the Okazaki fragments.

Leading strand - a strand of the DNA molecule which has been synthesized continuously.

Leathal - a defective gene which causes a death, if dominant - in the heterozygous state, if recessive - only in a case when both alleles in the same locus are defective.

Locus (loci) - a place on chromosome where a given gene resides.

Meiosis - two consecutive divisions of a nucleus after only one replication of chromosomes resulting in producing four haploid products (gametes).

Mendelian population - large interbreeding, panmictic population sharing a common genetic pool.

Missense mutation - substitution which changes the codon for the other one coding different amino acid.

Mutation - any change of the DNA sequence.

Nonsense mutation - substitution which changes the codon coding for amino acid for the other one coding for stop translation.

Outbreeding depression - reduction of the reproduction potential due to mating between genetically distant partners.

Panmictic populations - populations without any mating preferences or restrictions, random mating.

Parity rule I - deterministic rule. See Chargaff's rule 1.

Parity rule II - stochastic rule stating that the number of adenine in one DNA strand equals the number of thymine, the number of guanine equals the number of cytosine.

Phenotype - the observable traits of an individual resulting from the interaction between its genotype and the environmental demands.

Purifying selection - elimination of the defective alleles from the genetic pool by selection.

Recombination - any exchange of genetic information between two genetic units or inside one unit (chromosome), in particular - crossover.

Recessive (allele) - an allele, function of which is not seen in the heterozygous state, it is masked by dominant allele.

Replication - DNA replication, synthesis of a new DNA molecule on the template of the parental molecule.

Reversion - mutation which restores the function of a gene destroyed (or changed) by the previous mutation.

Sex chromosomes - chromosomes which are different in the two sexes.

Silent mutation - (synonymous mutation) substitution which changes the codon for a synonymous one, coding for the same amino acid.

Speciation - emerging (formation) of a new species.

Species - a group of interbreeding individuals which are reproductively isolated from individuals belonging to other species.

Stop codon - one of three codons meaning "stop translation".

Substitution matrix - a table of frequencies of substitutions (4 transitions and 8 transversions) characteristic for a given genome (usually different for the leading and the lagging DNA strands). S.m. is used for description of the mutational pressure.

Sympatric speciation - emerging of a new species inside another species, without any biological, physical or geographical barriers.

Synonymous mutation - see silent mutation.

Telomere - it is a region of repetitive DNA at the end of chromosomes, which protects the end of the chromosome from destruction.

Transcription - synthesis of the RNA molecule (single stranded) on template of the DNA molecule.

Transition - it is a kind of mutation - substitution of one purine (A or G) by the other one or substitution of one pyrimidine (C or T) by the other one.

Translation - synthesis of the protein (amino acid sequence) on the matrix of the mRNA molecule.

Translocation - transfer of a fragment of DNA from one position to the other one.

Transversion - it is a kind of mutation - substitution of purine (A or G) by pyrimidine (C or T) or vice versa.

Prepared by
Stanisław Cebrat

Author Index

Subject Index

Photos